Lecture Notes in Biomathematics

Managing Editor: S. Levin

15

Thomas Nagylaki

Selection in One- and Two-Locus Systems

Springer-Verlag
Berlin · Heidelberg · New York 1977

Author
Prof. Dr. Thomas Nagylaki
Dept. of Biophysics and Theoretical Biology
The University of Chicago
920 East 58th Street
Chicago, Illinois 60637/USA

Library of Congress Cataloging in Publication Data

Nagylaki, Thomas, 1944-
Selection in one- and two-locus systems.

(Lecture notes in biomathematics ; 15)
Bibliography: p.
Includes index.
1. Population genetics--Mathematical models.
2. Natural selection--Mathematical models. I. Title.
II. Series.
QH455.N33 575.1 77-8295

AMS Subject Classifications (1970): 92-01, 92-02, 92A05, 92A10.

ISBN 3-540-08247-6 Springer-Verlag Berlin · Heidelberg · New York
ISBN 0-387-08247-6 Springer-Verlag New York · Heidelberg · Berlin

Printing and binding: Beltz Offsetdruck, Hemsbach/Bergstr.
2145/3140-543210

PREFACE

Most of these notes were presented as part of a two-quarter course on theoretical population genetics at The University of Chicago. Almost all the students were either undergraduates in mathematics or graduate students in the biological sciences. The only prerequisites were calculus and matrices. As is done in these notes, biological background and additional mathematical techniques were covered when they were required. I have included the relevant problems assigned in the course.

My aim in these notes is to formulate the various models fairly generally, making the biological assumptions quite explicit, and to perform the analyses relatively rigorously. I hope the choice and treatment of topics will enable the reader to understand and evaluate detailed analyses of specific models and applications in the literature. No attempt has been made to review the literature or to assign credit. Most of the references are to papers directly germane to the subjects and approaches covered here. Frequency of reference is not intended to reflect proportionate contribution.

I am very grateful to Professor James F. Crow for helpful comments and to Mrs. Adelaide Jaffe for her excellent typing. I thank the National Science Foundation for its support (Grant No. DEB76-01550).

June 1976 Thomas Nagylaki

CONTENTS

1. INTRODUCTION 1

2. ASEXUAL HAPLOID POPULATIONS 5

2.1 *Selection* 5

2.2 *Mutation and Selection* 9

2.3 *Migration and Selection* 14

2.4 *Continuous Model with Overlapping Generations* 14

2.5 *Problems* 30

3. PANMICTIC POPULATIONS 33

3.1 *The Hardy-Weinberg Law* 33

3.2 *X-Linkage* 40

3.3 *Two Loci* 42

3.4 *Population Subdivision* 46

3.5 *Problems* 47

4. SELECTION AT AN AUTOSOMAL LOCUS 51

4.1 *Formulation for Multiple Alleles* 51

4.2 *Dynamics with Two Alleles* 55

4.3 *Dynamics with Multiple Alleles* 60

4.4 *Two Alleles with Inbreeding* 66

4.5 *Variable Environments* 68

4.6 *Intra-Family Selection* 71

4.7 *Maternal Inheritance* 73

4.8 *Meiotic Drive* 74

4.9 *Mutation and Selection* 75

4.10 *Continuous Model with Overlapping Generations* 79

4.11 *Problems* 92

5. NONRANDOM MATING 95

5.1 *Selfing with Selection* 96

5.2 *Assortative Mating with Multiple Alleles and Distinguishable Genotypes* 101

5.3 *Assortative Mating with Two Alleles and Complete Dominance* 102

5.4 *Random Mating with Differential Fertility* 107

5.5 *Self-Incompatibility Alleles* 112

5.6 *Pollen and Zygote Elimination* 115

5.7 *Problems* 122

6. MIGRATION AND SELECTION 124

6.1 *The Island Model* 124

6.2 *General Analysis* 130

6.3 *The Levene Model* 142

6.4 *Two Diallelic Niches* 146

6.5 *Problems* 150

7. X-LINKAGE 151

7.1 *Formulation for Multiallelic Selection and Mutation* . 151

7.2 *Selection with Two Alleles* 157

7.3 *Mutation-Selection Balance* 160

7.4 *Problems* 163

8. TWO LOCI 165

8.1 *General Formulation for Multiple Loci* 165

8.2 *Analysis for Two Multiallelic Loci* 167

8.3 *Two Diallelic Loci* 177

8.4 *Continuous Model with Overlapping Generations* 182

8.5 *Problems* 189

REFERENCES 191

SUBJECT INDEX 199

1. INTRODUCTION

Population genetics concerns the genetic structure and evolution of natural populations. The genetic composition of a population is usually described by genotypic proportions, which may depend on space and time. These genotypic frequencies are determined by a few elementary genetic principles and the following evolutionary factors.

Various genotypes may have different probabilities of surviving to adulthood and may reproduce at different rates. Differential mortality and fertility are the components of *selection*. Unless the population is in equilibrium, selection will change the genotypic and allelic frequencies in accordance with the expected number of progeny, called fitness, of the various genotypes. Natural selection has been recognized since Darwin as the directive force of adaptive evolution.

The action of selection is strongly affected by the *mating system*. If mating occurs without regard to the genotypes under consideration, we say it is random. This is the simplest situation and, at least approximately, appears to be frequently realized in nature. We say there is inbreeding if related individuals are more likely to mate than randomly chosen ones. Assortative mating refers to the tendency of individuals resembling each other with respect to the trait in question to mate with each other. Disassortative mating means that phenotypically dissimilar individuals mate more often than randomly chosen ones. Nonrandom mating influences genotypic frequencies. In the absence of selection, inbreeding does not change gene frequencies, but assortative and disassortative mating may. This will happen if the mating pattern is such that some genotypes have a higher probability of mating than others.

Mutation designates the change from one allelic form to another. Clearly, it directly alters gene frequencies.

In spatially structured populations, *migration* must be taken into account. It can affect not only the geographical composition of the population, but the amount of genetic variability as well.

Unless some of the parameters, such as the selection intensities, required to specify the elements of evolution described above fluctuate at random, the evolutionary forces will be deterministic. In a finite population, however, allelic frequencies will vary probabilistically due to the random sampling of genes from one generation to the next. This process is called *random genetic drift.* Its causes are (nonselective) random variation in the number of offspring of different individuals and the stochastic nature of Mendel's Law of Segregation. Evidently, the smaller the population, the larger is the evolutionary rôle of random drift. No matter how large the population is, however, the fate of rare genes still depends strongly on random sampling.

All three current views of evolution identify mutation as the source of raw material for evolutionary change, but differ in the emphasis placed on the other factors.

The main underpinning of Fisher's theory is his Fundamental Theorem of Natural Selection--that the rate of change of the mean fitness of a population is equal to the additive component of its genetic variance in fitness (Fisher, 1930). Fisher held that evolution occurs primarily by the deterministic increase in fitness of large populations under the action of natural selection. We may imagine the mean fitness as a surface in the space of suitable dynamical variables like gene frequencies, and consequently envisage that the population is climbing a hill on this surface. In Fisher's picture, random drift is responsible only for small chance fluctuations in the trajectory of the population. These notes are essentially an exposition of the theory underlying Fisher's view at the level of one and two loci.

Wright (1931, 1970) stressed that, due to multiple effects (pleiotropy) and interactions (epistasis) of loci and selection for an intermediate optimum, the fitness surface has many selective peaks. Small populations can "test" this surface by random drift, sometimes crossing a saddle from a lower selective hill to a higher one. He pointed out that if a species is divided into many such small populations, which exchange relatively few migrants, dispersion of selectively favored individuals may enable it to reach the highest peak on the surface. A comprehensive analytical or numerical treatment of this complex theory awaits future research.

The neutral theory of Kimura (1968) and of King and Jukes (1969) ascribes much of evolution, especially at the molecular level, to mutation and random drift. In its strongest form, this theory attributes a lot of the variation even in morphological characters to these two forces (Nei, 1975, pp.251-253). This is still not inconsistent with the fact that natural selection determines the nature of adaptation. The weakest form of the theory holds only that most amino acid substitutions are neutral. The neutral theory is mathematically quite highly developed.

It may be helpful to list with suitable general references the major aspects of theoretical population genetics not covered in these notes. Inbreeding and quantitative genetics do not require advanced mathematics and are discussed by Crow and Kimura (1970). The other topics require considerable use of probability, analysis, and differential equations. For random fluctuations in selection intensities, the reader may consult Karlin and Lieberman (1974) and Karlin and Levikson (1974). Geographical variation is discussed in Nagylaki (1977). Various aspects of random drift are treated by Moran (1962), Ewens (1969), Crow and Kimura (1970), and Kimura and Ohta (1971). Felsenstein and Taylor (1974) have compiled a bibliography of theoretical population genetics with almost complete coverage through the autumn of 1973.

The next chapter concerns selection in asexual haploid populations. Its purpose is to formulate and analyze in a simple setting many of the problems dealt with in these notes. Having obtained this perspective, the reader will be better prepared for the treatment of selection with the additional complexity of mating and Mendelian segregation and recombination in the remainder of the notes. Chapter 3 examines the structure of a randomly mating population in the absence of selection. The reader who already has a thorough understanding of Mendelism (including sex-linkage and recombination), which will be assumed in the subsequent chapters, may omit this chapter. The basic theory of selection at an autosomal locus is developed in Chapter 4. A few less common types of selection and modes of inheritance are discussed chiefly to provide more practice in formulating models. The remaining chapters are logically independent of each other.

We shall now give a minimum of very much simplified genetic background. More information will be introduced when it is required. Cavalli-Sforza and Bodmer (1971) present a good summary of the pertinent genetic background. Crow's (1976) lucid book is more detailed, but still quite concise.

The genetic material, deoxyribonucleic acid (DNA), consists of four bases, adenine (A), guanine (G), thymine (T), and cytosine (C), each of which is linked to a sugar and a phosphate, forming a *nucleotide*. These nucleotides are arranged in a double helix, the only possible pairings being A-T and G-C. Therefore, the information is in the sequence along a single helix. Three bases code for an amino acid, of which there are 20. A protein consists of at least several hundred amino acids. Roughly, the region of DNA determining a protein is a *locus* or *gene*. A particular sequence there is an *allele*. In population genetics, however, "gene" is sometimes used in the sense of "allele", as defined above (e.g., in "gene frequency"). A *chromosome* is composed of proteins and a single thread-like molecule of DNA, along which genes are arranged linearly. The number of different chromosomes is characteristic of each species. If the chromosomes in a set are single, the organism is called *haploid*; if they are doubled, the organism is *diploid*. Generally, bacteria, algae, mosses, and fungi are examples of haploid organisms, while higher plants and animals are diploid. We shall not consider polyploids, in which chromosomes are at least tripled. Very crudely, asexual haploids just duplicate themselves. At *meiosis*, in diploids the chromosomes separate, and each *gamete* (sperm or egg) carries a single set of chromosomes. The sets from a sperm and egg unite at *fertilization* to form a diploid *zygote*, from which the individual develops. In plants, pollen fertilize ova to form seeds.

2. ASEXUAL HAPLOID POPULATIONS

The study of haploid populations in this chapter will enable us to formulate and analyze many of the problems which concern us without the additional complication of mating and Mendelian segregation and recombination. If alleles are interpreted as genotypes, and mutation rates refer to zygotes rather than alleles, the formalism applies, regardless of ploidy, also to asexual species like the dandelion. As explained in Chapter 3, it applies also to a *Y*-linked locus in sexually-reproducing diploids. We shall expound the basic selection model with discrete nonoverlapping generations in Section 2.1, include mutation and migration in Sections 2.2 and 2.3, and treat overlapping generations with continuous time in Section 2.4.

2.1 *Selection*

We consider a single locus with alleles A_i, $i = 1, 2, \ldots, k$, and assume generations are discrete and nonoverlapping. Thus, the adults are replaced by their offspring in each generation. Although this assumption will hold for some laboratory populations, for natural populations of haploids it should be viewed as a simple approximation. Let the number of offspring carrying A_i in generation t, where $t = 0, 1, 2 \ldots,$ be $n_i(t)$. The total number of offspring,

$$N = \sum_i n_i, \tag{2.1}$$

must be sufficiently large to allow us to neglect random drift.

Let v_i designate the probability that an A_i offspring survives to reproductive age. The average number of progeny of an A_i adult is f_i. The *viabilities* v_i and *fertilities* f_i may be functions of the time t and the vector of population numbers, denoted by $\underset{\sim}{n}(t)$. The product $w_i = v_i f_i$ represents the *fitness* of an A_i individual. The $v_i n_i$ A_i adults in generation t contribute $f_i(v_i n_i) = w_i n_i$ A_i offspring to generation t+1. Therefore, the basic recursion relations

$$n_i(t+1) = w_i[t,\underset{\sim}{n}(t)]n_i(t) \tag{2.2}$$

depend only on the fitnesses, and not on the viabilities and fertilities separately. Since w_i is the expected number of progeny of an A_i juvenile, this is not surprising. The fundamental difference equations (2.2) determine $\underset{\sim}{n}(t)$ iteratively in terms of $\underset{\sim}{n}(0)$, provided the fitnesses $w_i(t,\underset{\sim}{n})$ are specified.

If A_i is lethal, $v_i = 0$, or causes sterility, $f_i = 0$, then $w_i = 0$. Otherwise, the w_i will usually not differ from each other by more than a few percent. If the population size is approximately constant, the average of the w_i will be close to unity. The numbers w_i-1 are called *selection coefficients*.

The proportion or *frequency* of the allele A_i among offspring is

$$p_i = \frac{n_i}{N}. \tag{2.3}$$

Unless stated otherwise, a prime will always signify the next generation. Then (2.1), (2.2), and (2.3) yield

$$N' = \sum_i n_i' = \sum_i w_i\left(\frac{n_i}{N}\right)N = \overline{w}N, \tag{2.4}$$

where

$$\overline{w} = \sum_i w_i p_i \tag{2.5}$$

is the *mean fitness* of the population and gives its rate of growth. The mean fitness is of great conceptual and analytical importance in the theory of selection.

The gene frequencies satisfy the recursion relations

$$p_i' = \frac{n_i'}{N'} = \frac{w_i n_i}{\overline{w}N} = p_i\left(\frac{w_i}{\overline{w}}\right). \tag{2.6}$$

We see at once from (2.5) and (2.6) that the relation

$$\sum_i p_i = 1 \tag{2.7}$$

holds for all time if initially true, as it must be. It is also apparent from (2.5) and (2.6) that the gene frequencies depend only on ratios of the fitnesses. All the w_i may be multipled by the same constant without altering the evolution of the allelic frequencies p_i. Exploiting this scale invariance often simplifies the algebra. We shall always employ a Δ to indicate the change in one generation. Thus, (2.6) gives

$$\Delta p_i = p_i'-p_i = \frac{p_i(w_i-\overline{w})}{\overline{w}}. \tag{2.8}$$

If the fitnesses are functions only of time, it is useful to iterate (2.2):

$$n_i(t) = n_i(0) \prod_{\tau=0}^{t-1} w_i(\tau). \tag{2.9}$$

Then the gene frequencies read

$$p_i(t) = \frac{p_i(0) \prod_{\tau=0}^{t-1} w_i(\tau)}{\sum_j p_j(0) \prod_{\tau=0}^{t-1} w_j(\tau)}. \tag{2.10}$$

It is often assumed that the fitnesses are constant, meaning, really, that they vary much more slowly than the other pertinent evolutionary parameters. In that case, (2.9) and (2.10) reduce to

$$n_i(t) = n_i(0)w_i^t, \tag{2.11}$$

$$p_i(t) = \frac{p_i(0)w_i^t}{\sum_j p_j(0)w_j^t}. \tag{2.12}$$

Suppose A_1 is the fittest allele: $w_1 > w_i$, $i > 1$. Equation (2.12) informs us immediately that $p_1(t) \to 1$ as $t \to \infty$. Since the population number can be changed each generation without changing gene frequencies, this means that the population size will remain finite, and the fittest gene will be ultimately *fixed*, the others being *lost*. Of course, all statements of this sort presuppose the allele under consideration is initially present in the population. For instance, here we posit $p_1(0) > 0$.

It will be helpful to distinguish three levels of description of the evolution of a population. We shall refer to the specification of the variables of interest as functions of time as a *complete solution*. Equation (2.12) is an example. Often, even though one cannot obtain a complete solution, one can determine the fate of the population for all initial conditions. We may call this a *complete* or *global analysis*. The statement that with constant fitnesses the fittest allele is fixed, falls in this category. If we cannot carry out a complete analysis, we may still obtain some information of evolutionary interest by locating all the equilibria of the system and investigating its behavior in the neighborhood of these stationary states. In the problem treated above, a part of such a *local analysis* would be to observe that $p_1 = 1$ is an equilibrium, and to show that if $p_1(0)$ is sufficiently close to 1, then $p_1(t) \to 1$ as $t \to \infty$.

We characterize the local stability of equilibria as follows. If, in terms of a suitable metric for the variables of the problem, the population will remain within an arbitrarily small preassigned distance of the equilibrium, provided it starts sufficiently close to the equilibrium, we say the equilibrium is *stable*. Otherwise, it is *unstable*. An equilibrium is *asymptotically stable* if it is stable, and a population starting sufficiently close to the equilibrium converges to it. The gene frequency equilibrium $p_1 = 1$ is globally asymptotically stable, while the equilibrium $p_2 = 1$ is unstable. If $w_i = 1$ for all i, every point is a stable equilibrium, but the stability is not asymptotic. The word "asymptotic" is frequently omitted in population genetics.

We proceed now to study the change in mean fitness. From (2.5) we have

$$\begin{aligned} \Delta\bar{w} &= \sum_i (p_i'w_i'-p_iw_i) \\ &= \sum_i [p'(\Delta w_i+w_i)-p_iw_i] \\ &= \overline{\Delta w} + \sum_i w_i\Delta p_i, \end{aligned} \tag{2.13}$$

where

$$\overline{\Delta w} = \sum_i p_i'\Delta w_i \tag{2.14}$$

is the mean of the fitness changes over the next generation. Substituting (2.8) into (2.13), we find

$$\Delta\bar{w} = \overline{\Delta w} + \bar{w}^{-1} \sum_i p_iw_i(w_i-\bar{w}). \tag{2.15}$$

But (2.5) informs us that

$$\sum_i p_i(w_i-\bar{w}) = 0,$$

so we may subtract $\bar{w}$ from the first w_i in (2.15) to obtain

$$\Delta\bar{w} = \overline{\Delta w} + \bar{w}^{-1}V, \tag{2.16}$$

where

$$V = \sum_i p_i(w_i-\bar{w})^2 \tag{2.17}$$

is the genic variance in fitness. The simple steps in this derivation are often useful.

Equation (2.16) is a simple case of Fisher's Fundamental Theorem of Natural Selection (Fisher, 1930). With constant fitnesses, $\overline{\Delta w} = 0$, so $\Delta\bar{w} = \bar{w}^{-1}V \geq 0$, i.e., the mean fitness is nondecreasing. Since $V = 0$ if and only if $p_i = 0$ or $w_i = \bar{w}$ for all i, (2.6) implies $\Delta\bar{w} = 0$ only at equilibrium. Thus, selection increases the mean fitness, using up the genic variance. If the selection coefficients are small, we may choose all the fitnesses to be close to 1. Then $\bar{w}$ will be approximately unity, and the rate of change of the mean fitness will be roughly equal to the genic variance.

With only two alleles, it is customary to put $p = p_1$ and $q = p_2 = 1-p$. From (2.8) we deduce

$$\Delta p = pq(w_1-w_2)/\bar{w}, \tag{2.18}$$

and the variance in fitness reduces to

$$V = pq(w_1-w_2)^2. \tag{2.19}$$

2.2 *Mutation and Selection*

Let us consider first mutation without selection. We designate the probability that an A_i individual has an A_j offspring for $i \neq j$ by the mutation rate u_{ij}. It will be convenient to use the convention $u_{ii} = 0$ for all i. Generally, mutation rates are quite small; 10^{-6} is a representative value. Mutation rates at the nucleotide level are on the order of 10^{-10}. The gene frequency change in one generation clearly reads

$$\Delta p_i = \sum_j p_j u_{ji} - p_i \sum_j u_{ij}. \tag{2.20}$$

By interchanging dummy variables in one of the sums, often a useful device, we observe directly that

$$\sum_i \Delta p_i = 0,$$

so that (2.7) is preserved. Since the total mutation rate must not exceed unity, therefore $\Delta p_i \geq -p_i$, whence $p_i' \geq 0$, as required.

For two alleles, one commonly writes $u = u_{12}$ and $v = u_{21}$. With $p = p_1$, as above, (2.20) becomes

$$\Delta p = v-(u+v)p. \tag{2.21}$$

At equilibrium, $\Delta p = 0$, so the frequency of A_1 is

$$\hat{p} = \frac{v}{u+v}. \tag{2.22}$$

We shall follow convention and indicate equilibrium values by a caret. Equilibria like (2.22), with more than one allele present, are called *polymorphic*. As expected, if mutation is irreversible, i.e., $u = 0$ or $v = 0$, the allele whose frequency is decreasing is absent at equilibrium.

It is frequently convenient to study the deviation from equilibrium. Substituting $x = p-\hat{p}$ into (2.21), we find

$$x' = (1-u-v)x,$$

with the solution

$$x(t) = x(0)(1-u-v)^t. \tag{2.23}$$

Therefore, there is global convergence to (2.22) at the geometric rate $1-u-v$. Note that the time required for significant gene frequency change,

$$t = \frac{\ln[x(t)/x(0)]}{\ln(1-u-v)} \approx \frac{\ln[x(0)/x(t)]}{u+v},$$

is very long, typically about 10^6 generations.

It is often useful to approximate powers like the one in (2.23) by exponentials. For $|\varepsilon| \ll 1$ and $\varepsilon^2 t \ll 1$, $(1-\varepsilon)^t = \exp[t\ln(1-\varepsilon)] = \exp[-t(\varepsilon+\frac{1}{2}\varepsilon^2+\ldots)] \approx e^{-\varepsilon t}$. Thus, we may rewrite (2.23) as

$$x(t) \approx x(0)e^{-(u+v)t}. \tag{2.24}$$

The fact that (2.24) becomes inaccurate as $(u+v)^2 t$ approaches 1 is irrelevant because by that time $x(t)$ is extremely close to 0.

To include selection, we set up the formal scheme

$$\begin{array}{ccccccc} \text{Offspring} & \xrightarrow{\hspace{2em}} & \text{Adult} & \xrightarrow{\hspace{2em}} & \text{Offspring} & \xrightarrow{\hspace{2em}} & \text{Offspring} \\ & \text{viability} & & \text{fertility} & & \text{mutation} & \\ p_i & & & & p_i^* & & p_i' \end{array}$$

with the indicated gene frequencies. Let R_{ij} be the probability that a gamete from an A_i offspring carries A_j. Recalling (2.6), we have

$$p_i^* = p_i\left(\frac{w_i}{\bar{w}}\right), \tag{2.25a}$$

$$p_i' = \sum_j p_j^* R_{ji}, \tag{2.25b}$$

where $\bar{w}$ is still given by (2.5). It is important to note that (2.25) correctly describes the biological situation that, while selection acts on the phenotype, which develops from the offspring genotype, the germ cells mutate with no phenotypic effect at rates u_{ij}, related to R_{ij} by

$$R_{ij} = \delta_{ij}\left(1 - \sum_k u_{ik}\right) + u_{ij}. \tag{2.26}$$

The Kronecker delta, δ_{ij}, is defined by $\delta_{ij} = 1$ if $i = j$ and $\delta_{ij} = 0$ if $i \neq j$.

From (2.25b) and (2.26) we derive

$$\Delta p_i = p_i' - p_i = p_i^* - p_i + \sum_j p_j^* u_{ji} - p_i^* \sum_j u_{ij}.$$

If selection is weak, since $u_{ij} \ll 1$, we may neglect $(p_i^* - p_i)u_{ij}$ for all i and j to obtain

$$\Delta p_i \approx \Delta p_i(\text{selection}) + \Delta p_i(\text{mutation}),$$

where

$$\Delta p_i(\text{selection}) = p_i^* - p_i,$$

$$\Delta p_i(\text{mutation}) = \sum_j p_j u_{ji} - p_i \sum_j u_{ij}.$$

Let us analyze the diallelic case. Choosing, without loss of generality, $w_1 = 1$, $w_2 = 1+s$, $s > 0$, in the notation introduced above, (2.25) reduces to the *linear fractional transformation*

$$p' = \frac{\alpha + \beta p}{\gamma + \delta p} \tag{2.27}$$

with $\alpha = v(1+s)$, $\beta = 1-sv-u-v$, $\gamma = 1+s$, $\delta = -s$. Since (2.27) occurs in several models, we shall discuss it for arbitrary values of its parameters. The two solutions of $p' = p$ are

$$p_{\pm} = (2\delta)^{-1}(\beta-\gamma\pm Q^{1/2}), \tag{2.28}$$

where

$$Q = (\beta-\gamma)^2 + 4\alpha\delta. \tag{2.29}$$

The trivial case $\delta = 0$ corresponds to (2.21), so we suppose $\delta \neq 0$. We also assume $\alpha\delta \neq \beta\gamma$, for otherwise (2.27) shows that $p' = \alpha/\gamma$.

If $Q \neq 0$, then $p_+ \neq p_-$, hence

$$y = \frac{p-p_+}{p-p_-} \tag{2.30}$$

satisfies $y' = \lambda y$, where

$$\lambda = \frac{\beta+\gamma-Q^{1/2}}{\beta+\gamma+Q^{1/2}}. \tag{2.31}$$

Therefore,

$$y(t) = y(0)\lambda^t, \tag{2.32}$$

and (2.27) has the solution

$$p(t) = \frac{p_- y(t)-p_+}{y(t)-1}. \tag{2.33}$$

Of course, $y(0)$ is evaluated from (2.30). There are two cases.

1. $Q < 0$: Here p_+ and p_- are complex. Since $|\lambda| = 1$, (2.32) has the form

$$y(t) = y(0)e^{-i\theta t}, \tag{2.34}$$

where

$$\theta = 2\tan^{-1}\left[\frac{(-Q)^{1/2}}{\beta+\gamma}\right]. \tag{2.35}$$

2. $Q > 0$: Now p_+ and p_- are real, and there are three subcases.

   (a) $\beta+\gamma > 0$: From (2.31) we see that $|\lambda| < 1$, whence $y(t) \to 0$ as $t \to \infty$. Therefore, (2.33) implies that $p(t) \to p_+$ as $t \to \infty$.

   (b) $\beta+\gamma < 0$: Since $|\lambda| > 1$, therefore, $y(t) \to \infty$, and consequently $p(t) \to p_-$ as $t \to \infty$.

   (c) $\beta+\gamma = 0$: Here $\lambda = -1$, and hence $y(t) = y(0)(-1)^t$. Therefore, $p(t)$ alternates between $p(0)$ and $p(1)$. This also holds for $Q < 0$, for then (2.35) yields $\theta = \pi$.

It remains to analyze the case with $Q = 0$.

3. $Q = 0$: Equation (2.28) tells us that there is a single equilibrium

$$\hat{p} = (2\delta)^{-1}(\beta-\gamma). \tag{2.36}$$

We find that

$$z = (p-\hat{p})^{-1} \tag{2.37}$$

satisfies, for $\beta+\gamma \neq 0$,

$$z' = z+\frac{2\delta}{\beta+\gamma}, \tag{2.38}$$

with the obvious solution

$$z(t) = z(0)+\frac{2\delta t}{\beta+\gamma}. \tag{2.39}$$

Equations (2.37) and (2.39) inform us that

$$p(t) \sim \hat{p}+\frac{\beta+\gamma}{2\delta t} \tag{2.40}$$

as $t \to \infty$. Thus, the ultimate rate of convergence to $\hat{p}$ is algebraic. If $\beta+\gamma = 0$, (2.29) tells us that $\alpha\delta = \beta\gamma$, so that $p' = \alpha/\gamma$.

In genetic problems, we shall be concerned with the mapping of some interval $I:[a,b]$ into itself. Then we can restrict the possible equilibrium structures for any continuous map $p' = f(p)$. Suppose a and b are not equilibria: $f(a) \neq a$ and $f(b) \neq b$. Hence, $f(a) > a$ and $f(b) < b$, so $g(p) = f(p)-p$ must change sign an odd number of times in I. Since f maps I into itself, therefore $g(p)$ is finite for p in I, and hence, counting multiplicity, has an odd number of zeroes in I. Thus, still counting multiplicity, f has an odd number of fixed points in I. For (2.27), provided $p_\pm \neq 0,1$, this means cases 1 and 3 may be excluded, and in case 2 either p_+ or p_- is in I, but not both.

Let us apply the theory just developed to our mutation-selection problem. It is easy to see that $Q \geq 0$, equality holding if and only if $u = 0$ and $s = v(1+s)$. Since usually $s \gg v$, we assume $Q > 0$. Furthermore, it is trivial to verify that $\beta+\gamma \geq 0$, with equality only in the unbiological situation $u = v = 1$. Therefore, $p(t)$ converges to p_+ globally. It follows that $0 \leq p_+ \leq 1$. From the formula

$$p_\pm = (2s)^{-1}\{s+sv+u+v\mp[(s+sv+u+v)^2-4sv(1+s)]^{1/2}\}, \tag{2.41}$$

with a bit of algebra one can show that $p_- > 1$. With the biologically trivial assumption $u+v \leq 1$, one can prove from (2.31) that $\lambda \geq 0$, so that we conclude from (2.30) that the convergence is without oscillation. If $v = 0$, both selection and mutation decrease p, so we expect $p_+ = 0$. Indeed, (2.41) yields $p_+ = 0$ and $p_- = 1+us^{-1}$. With $u = 0$, (2.41) reduces to $p_+ = v(1+s)/s$, $p_- = 1$. In the biologically important case $u,v \ll 1,s$, linearizing (2.41) in u and v yields $p_+ \approx v(1+s)/s$ and $p_- \approx 1+us^{-1}$. As expected, the equilibrium frequency, $p_+ \ll 1$.

2.3 *Migration and Selection*

We assume a proportion m of the population is replaced each generation by migrants with fixed gene frequencies $\overline{p}_i$. More complicated migration-selection schemes than this island model will be discussed for diploids in Chapter 6. To write our recursion relations, we replace mutation by migration in the formal mutation-selection scheme of Section 2.2. Then (2.25b), which we may rewrite as

$$p_i' = p_i^* + \sum_j p_j^* u_{ji} - p_i^* \sum_j u_{ji}, \tag{2.42}$$

becomes

$$p_i' = p_i^* + m(\overline{p}_i - p_i^*). \tag{2.43}$$

But the substitution $u_{ij} = m\overline{p}_j$, $i \neq j$, reduces (2.42) to (2.43), showing that migration is a special case of mutation.

2.4 *Continuous Model with Overlapping Generations*

Our formulation will be based on that of Cornette (1975) for diploids. Time, measured in arbitrary units, flows continuously. Let $\nu_i(t,x)\Delta x$ be the number of A_i individuals between the ages of x and $x+\Delta x$ at time t. The total number of A_i individuals at time t is

$$n_i(t) = \int_0^\infty \nu_i(t,x)dx. \tag{2.44}$$

If no individual survives beyond age X, then $\nu_i(t,x) = 0$ for $x > X$. The total population size is

$$N(t) = \sum_i n_i(t). \tag{2.45}$$

We set up equations for our fundamental variables, $\nu_i(t,x)$, with the aid of the life table, $l_i(t,x)$. We start observing the population at time $t = 0$. For $t \geq x$, $l_i(t,x)$ represents the probability that an A_i individual born at time $t-x$ survives to age x. If $t < x$, $l_i(t,x)$ designates the probability that an A_i individual aged $x-t$ at time 0 survives to age x. By definition

$$l_i(t,0) = 1 \quad \text{and} \quad l_i(0,x) = 1. \tag{2.46}$$

The number of A_i individuals born from time $t-\Delta x$ to time t is the same as the number of A_i individuals between ages 0 and Δx at time t, $\nu_i(t,0)\Delta x$. This assertion neglects the mortality of the newborn since that is proportional to $(\Delta x)^2$. It follows that the number of A_i individuals born per unit time at time t is $\nu_i(t,0)$. For $t \geq x$, we equate the number of A_i individuals aged x at time t to the number of births at time $t-x$ times the proportion surviving to age x. For $t < x$, we replace the number of births at time $t-x$ by the number aged $x-t$ at time 0. We conclude

$$\nu_i(t,x) = \begin{cases} \nu_i(t-x,0)\,l_i(t,x), & t \geq x, \\ \nu_i(0,x-t)\,l_i(t,x), & t < x. \end{cases} \tag{2.47}$$

Thus, we have expressed the age distribution in terms of its initial value, the birth rate, and the life table. In principle, the birth rate, $\nu_i(t,0)$, may be calculated from the integral equation given in Problem 2.10.

Let us demonstrate next that the life table is completely determined by the mortality. We designate the probability that an A_i individual aged x at time t dies between ages x and $x+\Delta x$ by $d_i(t,x)\Delta x$. Then

$$l_i(t,x)-l_i(t+\Delta x,x+\Delta x) = l_i(t,x)d_i(t,x)\Delta x+O[(\Delta x)^2], \tag{2.48}$$

where $O[(\Delta x)^2]$ refers to terms proportional to $(\Delta x)^2$, or smaller, as $\Delta x \to 0$. We rewrite (2.48) as

$$\frac{l_i(t+\Delta x,x+\Delta x)-l_i(t,x+\Delta x)}{\Delta x}+\frac{l_i(t,x+\Delta x)-l_i(t,x)}{\Delta x} = -d_i(t,x)l_i(t,x)+O(\Delta x),$$

and let $\Delta x \to 0$ to obtain

$$\frac{\partial l_i}{\partial t}+\frac{\partial l_i}{\partial x} = -d_i(t,x)l_i(t,x). \tag{2.49}$$

To solve this partial differential equation, we put $y = \frac{1}{2}(t+x)$, $z = \frac{1}{2}(t-x)$, $D_i(y,z) = d_i(t,x)$, and $L_i(y,z) = \ln l_i(t,x)$. From (2.46) and (2.49) we deduce the problem

$$\frac{\partial L_i}{\partial y} = -D_i(y,z), \quad L_i(y,y) = 0, \quad L_i(y,-y) = 0. \tag{2.50}$$

Figure 2.1 shows that the solution of (2.50) in the first quadrant of the xt-plane reads

$$L_i(y,z) = \begin{cases} -\int_z^y D_i(y',z)dy', & z \geq 0, \\ -\int_{-z}^y D_i(y',z)dy', & z < 0. \end{cases} \tag{2.51}$$

Substituting $\xi = y'-z$ into (2.51) and returning to the original variables leads to

$$l_i(t,x) = \begin{cases} \exp\left[-\int_0^x d_i(\xi+t-x,\xi)d\xi\right], & t \geq x, \\ \exp\left[-\int_{x-t}^x d_i(\xi+t-x,\xi)d\xi\right], & t < x. \end{cases} \tag{2.52}$$

Returning to the age distribution, we observe that (2.48) holds with l_i replaced by ν_i, and hence so does (2.49):

$$\frac{\partial \nu_i}{\partial t} + \frac{\partial \nu_i}{\partial x} = -d_i(t,x)\nu_i(t,x). \tag{2.53}$$

Using (2.44) and (2.53), and noting that $\nu_i(t,\infty) = 0$, we calculate the time derivative

$$\dot{n}_i(t) = \int_0^\infty \frac{\partial \nu_i}{\partial t} dx$$

$$= -\int_0^\infty \left[\frac{\partial \nu_i}{\partial x} + d_i(t,x)\nu_i(t,x)\right] dx$$

$$= \nu_i(t,0) - \int_0^\infty d_i(t,x)\nu_i(t,x)dx. \tag{2.54}$$

We define $b_i(t,x)$ as the rate at which A_i individuals of age x at time t give birth. Then

$$\nu_i(t,0) = \int_0^\infty b_i(t,x)\nu_i(t,x)dx. \tag{2.55}$$

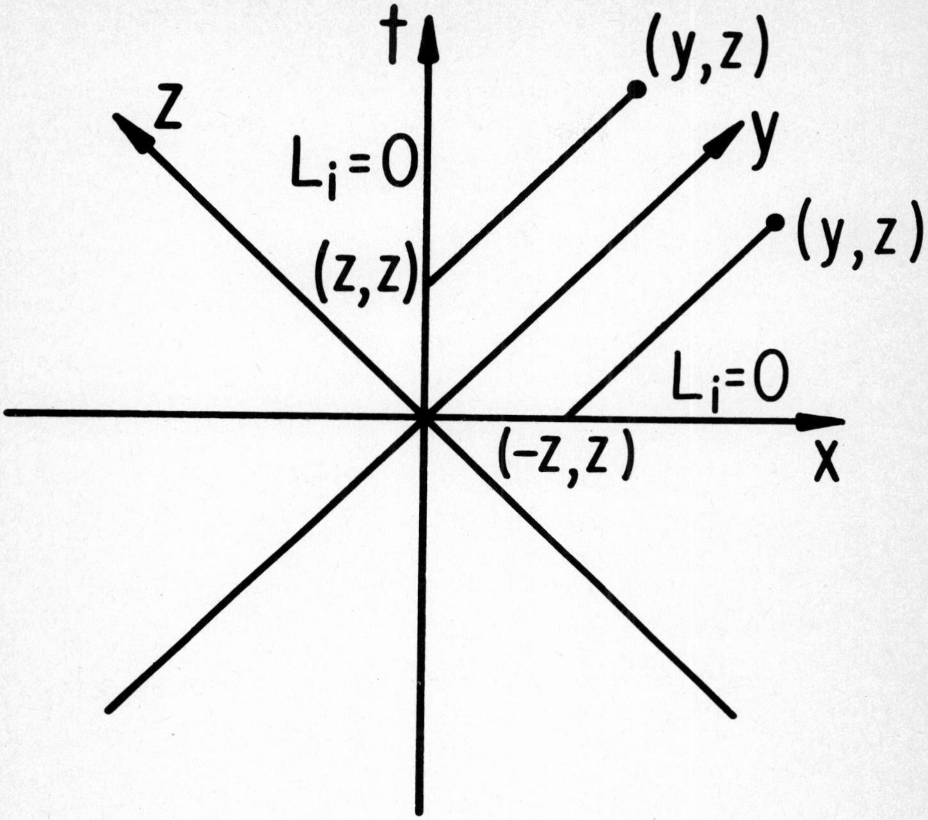

Fig. 2.1. The solution of Eq. (2.50)

Substituting (2.55) into (2.54) gives

$$\dot{n}_i(t) = \int_0^{\infty} m_i(t,x)\nu_i(t,x)dx, \tag{2.56}$$

where

$$m_i(t,x) = b_i(t,x)-d_i(t,x). \tag{2.57}$$

The usual model without age structure is obtained by assuming that $m_i(t,x) = m_i(t)$, independent of age. This will usually not happen unless the birth and death rates are each age-independent: $b_i(t,x) = b_i(t)$ and $d_i(t,x) = d_i(t)$. Note that with age-independent mortalities, setting $\tau = \xi+t-x$, (2.52) simplifies to

$$l_i(t,x) = \begin{cases} \exp\left[-\int_{t-x}^{t} d_i(\tau)d\tau\right], & t \geq x, \\ \\ \exp\left[-\int_{0}^{t} d_i(\tau)d\tau\right], & t < x. \end{cases} \quad (2.58)$$

Since the age at $t = 0$ is now irrelevant, $l_i(t,x)$ is independent of x for $t < x$. If mortality is also time-independent, $d_i(t) = d_i$, (2.58) yields an exponential life table depending on the time interval from the first observation of the individual to time t:

$$l_i(t,x) = \begin{cases} e^{-d_i x}, & t \geq x, \\ \\ e^{-d_i t}, & t < x. \end{cases} \quad (2.59)$$

Let us suppose that $m_i(t,x)$ is independent of x. This is exact for some populations like bacteria, but must be viewed as an approximation complementing the discrete nonoverlapping generations model for others. Then (2.56) shows that the Malthusian parameters $m_i[t,\underset{\sim}{n}(t)]$ determine the population numbers:

$$\dot{n}_i(t) = m_i[t,\underset{\sim}{n}(t)]n_i(t). \quad (2.60)$$

Due to the restrictive nature of the age-independence condition, it is important to observe that (2.60) also holds if the population has reached a stable age distribution. This may often be approximately true in natural populations. Let $\lambda_i(x)\Delta x$, independent of time by assumption, be the proportion of A_i individuals between ages x and $x+\Delta x$. Then $\nu_i(t,x) = n_i(t)\lambda_i(x)$,

$$\int_0^\infty \lambda_i(x)dx = 1,$$

and (2.56) yields

$$\dot{n}_i(t) = n_i(t) \int_0^\infty m_i(t,x)\lambda_i(x)dx.$$

With the expected identification of the Malthusian parameters, this has the form (2.60).

Equation (2.60) is the analogue of (2.2), and our exposition will follow Section 2.1. The gene frequencies $p_i(t)$ are still given by (2.3). From (2.3) and (2.60) we derive

$$\dot{N} = \bar{m}N, \tag{2.61}$$

where

$$\bar{m} = \sum_i m_i p_i \tag{2.62}$$

is the mean fitness in continuous time. Again using (2.3) and (2.60), we find that the gene frequencies satisfy

$$\dot{p}_i = p_i(m_i - \bar{m}). \tag{2.63}$$

As expected, $\dot{p}_i = 0$ if $p_i = 0$ and

$$\frac{d}{dt} \sum_i p_i = 0,$$

whence $p_i(0) \geq 0$ and $\sum_i p_i(0) = 1$ imply $p_i(t) \geq 0$ and $\sum_i p_i(t) = 1$, $t \geq 0$.

If $m_i(t,\underset{\sim}{n}) = m_i(t)$, (2.60) gives

$$n_i(t) = n_i(0) \exp\left[\int_0^t m_i(\tau)d\tau\right]. \tag{2.64}$$

Hence,

$$p_i(t) = \frac{p_i(0) \exp\left[\int_0^t m_i(\tau)d\tau\right]}{\sum_j p_j(0) \exp\left[\int_0^t m_j(\tau)d\tau\right]}. \tag{2.65}$$

With constant fitnesses m_i, (2.64) and (2.65) reduce to

$$n_i(t) = n_i(0)e^{m_i t}, \tag{2.66}$$

$$p_i(t) = \frac{p_i(0)e^{m_i t}}{\sum_j p_j(0)e^{m_j t}}. \tag{2.67}$$

In generations, (2.66) agrees with (2.11) if $w_i = e^{m_i}$. If A_1 is the fittest allele: $m_1 > m_i$, $i > 1$, (2.67) informs us that $p_1(t) \to 1$ as $t \to \infty$.

We shall now examine the rate of change of the mean fitness, $\dot{\bar{m}}$.

From (2.62) and (2.63) we obtain

$$\dot{\bar{m}} = \sum_i (\dot{m}_i p_i + m_i \dot{p}_i)$$

$$= \bar{\dot{m}} + \sum_i p_i m_i (m_i - \bar{m})$$

$$= \bar{\dot{m}} + V, \qquad (2.68)$$

where

$$\bar{\dot{m}} = \sum_i \dot{m}_i p_i \qquad (2.69)$$

is the mean rate of change of the fitnesses, and

$$V = \sum_i p_i (m_i - \bar{m})^2 \qquad (2.70)$$

represents the genic variance in fitness.

Equation (2.68) is the simplest continuous case of Fisher's Fundamental Theorem of Natural Selection (Fisher, 1930). With constant Malthusian parameters, $\bar{\dot{m}} = 0$, whence $\dot{\bar{m}} = V \geq 0$, with equality only at equilibrium. So, as for discrete nonoverlapping generations, selection increases the mean fitness at a rate equal to the genic variance, reducing the latter to zero in the process.

With only two alleles, (2.63) and (2.70) reduce to

$$\dot{p} = pq(m_1 - m_2), \qquad (2.71)$$

$$V = pq(m_1 - m_2)^2. \qquad (2.72)$$

As an example of a deteriorating environment, we take $m_i = r_i - at$ with r_i and a constant and $a > 0$. Since (2.63) yields $\dot{p}_i = p_i(r_i - \bar{r})$, the gene frequencies evolve as with $a = 0$, but (2.68) becomes $\dot{\bar{m}} = V - a$, where $V = \sum_i p_i (r_i - \bar{r})^2$. Therefore, $\dot{\bar{m}} < 0$ for sufficiently large a. Since we have already seen, and will find repeatedly, that the mean fitness is a most useful and illuminating function, this example leads us to seek a suitable generalization.

Assume there exists a function $F(\underset{\sim}{p}, t)$ with the following properties:

1. For all $\varepsilon > 0$, there exists $\eta > 0$ such that if the population deviates from gene frequency equilibrium (in some suitable metric)

by at least ε, then the time derivative of F along trajectories of the population satisfies $\dot{F}(\underset{\sim}{n},t) \geq \eta$.

2. F is bounded above.

Then the population converges to gene frequency equilibrium. For, if it did not, it would always be at least ε away from it, for some $\varepsilon > 0$, so there would exist some $\eta > 0$ such that $\dot{F} \geq \eta$. Therefore, $F(t) \geq \eta t - F(0) \to \infty$ as $t \to \infty$, contradicting condition 2.

To try to find F, it is easiest to look for a function satisfying $\dot{F} \geq 0$, with $\dot{F} = 0$ if and only if the population is in gene frequency equilibrium. If $\dot{F} = \dot{F}(\underset{\sim}{p})$, these requirements imply condition 1. If $F = F(\underset{\sim}{p})$, condition 2 holds. For instance, suppose

$$m_i = f_i(p_i) + g(\underset{\sim}{n},t). \tag{2.73}$$

From (2.69) we get

$$\dot{\bar{m}} = \sum_i p_i\left(\frac{df_i}{dp_i}\dot{p}_i + \frac{dg}{dt}\right).$$

We define

$$h(\underset{\sim}{p}) = \sum_i \int_0^{p_i} f_i'(x)x\,dx,$$

where the prime indicates the derivative here. Therefore

$$\dot{\bar{m}} = \frac{d}{dt}[h(\underset{\sim}{p}) + g(\underset{\sim}{n},t)],$$

whence (2.68) yields $\dot{F} = V(\underset{\sim}{p})$, with

$$\begin{aligned} F(\underset{\sim}{p}) &= \bar{m} - h(\underset{\sim}{p}) - g(\underset{\sim}{n},t) \\ &= \bar{f}(\underset{\sim}{p}) - h(\underset{\sim}{p}). \end{aligned} \tag{2.74}$$

Clearly, (2.73) subsumes constant fitnesses ($F = \bar{m}$) and $m_i = r_i - at$ ($F = \bar{r}$). We consider two less trivial examples.

1. Frequency dependence:

Rare alleles may be favored due to genotype-dependent resource requirements, special skills, etc. The simplest hypothesis is $m_i = r_i - a_i p_i$, where r_i and a_i are constant and $a_i > 0$. From (2.74) we calculate $F = \bar{r} - \frac{1}{2}\sum_i a_i p_i^2$. The diallelic case is easy to analyze.

From (2.71) we have

$$\dot{p} = pq[(r_1 - r_2 + a_2) - (a_1 + a_2)p]. \tag{2.75}$$

Since $p(t)$ changes continuously and time does not appear explicitly in (2.75), all the global results follow from the sign of $\dot{p}$. We find

(a) $r_1 - r_2 + a_2 \leq 0$: Evidently, $\dot{p} < 0$ for $0 < p < 1$. Therefore, $p(t) \to 0$ as $t \to \infty$.

(b) $r_1 - r_2 + a_2 > 0$: We separate two subcases.

(i) $r_1 - r_2 \geq a_1$: Now $\dot{p} > 0$ for $0 > p > 1$, hence $p(t) \to 1$.

(ii) $r_1 - r_2 < a_1$: This is the case of strong frequency dependence, i.e., large a_i. There is a polymorphic equilibrium

$$\hat{p} = \frac{r_1 - r_2 + a_2}{a_1 + a_2},$$

and $\operatorname{sgn} \dot{p} = \operatorname{sgn}(\hat{p} - p)$, where $\operatorname{sgn} x = \pm 1$, $x \gtrless 0$. Consequently, $p(t) \to \hat{p}$.

It is, of course, trivial to integrate (2.75) directly if more detailed information is required.

2. Population regulation:

We consider $m_i = r_i + g(\underset{\sim}{n})$, where the r_i are positive constants, and there exists N_0 such that $m_i < 0$ for $N > N_0$. According to (2.74), $F = \bar{r}$, and (2.63) yields $\dot{p}_i = p_i(r_i - \bar{r})$. Thus, this simple form of population regulation has no effect on gene frequencies. In the special case $g(\underset{\sim}{n}) = -\bar{r}N/K$, where K is a positive constant called the carrying capacity, from (2.61) we obtain

$$\dot{N} = \bar{r}N\left(1 - \frac{N}{K}\right), \tag{2.76}$$

which tells us that $N(t) \to K$ as $t \to \infty$. If the population is monomorphic, $\bar{r}$ is replaced by the constant r in (2.76) to give the logistic equation, with the solution

$$N(t) = K\left\{1 + \left[\frac{K - N(0)}{N(0)}\right]e^{-rt}\right\}^{-1},$$

sketched in Figure 2.2.

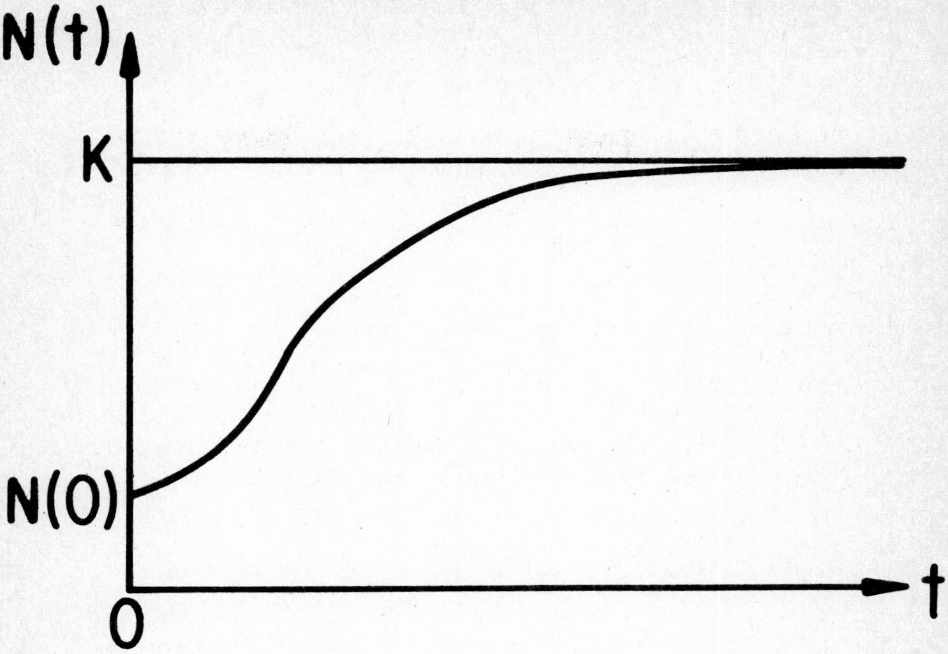

Fig. 2.2. The logistic curve.

As a final example, we shall discuss a model of competitive selection motivated by work of Kimura (1958), who analyzed the triallelic model with equal selection coefficients, and Nei (1971), who treated two alleles in discrete time. We assume that the genotypes compete in pairs, c_{ij} being the probability that A_i defeats A_j. Hence, $c_{ij} = 1 - c_{ji}$. We suppose a victorious individual gives birth, and a defeated one dies. Measuring time in generations, we posit Δt combats for each individual in time Δt. Therefore, the fertilities and mortalities are the probabilities of victory and defeat. Assuming pairings at random and putting $c_{ij} = \frac{1}{2}(1 + s_{ij})$, $s_{ij} = -s_{ji}$, for convenience, we have

$$b_i = \sum_j c_{ij} p_j = \frac{1}{2} + \frac{1}{2} \sum_j s_{ij} p_j,$$

$$d_i = \sum_j c_{ji} p_j = \frac{1}{2} - \frac{1}{2} \sum_j s_{ij} p_j.$$

Therefore, we obtain the frequency-dependent selection pattern

$$m_i = b_i - d_i = \sum_j s_{ij} p_j,$$

from which it follows immediately that $\bar{m} = 0$. Hence, the population size is constant. The allelic frequencies satisfy

$$\dot{p}_i = p_i \sum_j s_{ij} p_j. \tag{2.77}$$

The diallelic case is trivial: $\dot{p} = s_{12} pq$, so the frequency of the advantageous gene tends to unity.

For three alleles, we let $s_{12} = a$, $s_{23} = b$, $s_{31} = c$, and write out (2.77):

$$\dot{p}_1 = p_1(ap_2 - cp_3), \tag{2.78a}$$

$$\dot{p}_2 = p_2(-ap_1 + bp_3), \tag{2.78b}$$

$$\dot{p}_3 = p_3(cp_1 - bp_2). \tag{2.78c}$$

We exclude the trivial case $a = b = c = 0$, for then $p_i(t) = p_i(0)$. There are always three trivial equilibria, (1,0,0), (0,1,0), (0,0,1), corresponding to the fixation of A_1, A_2, A_3. If at least two of a,b,c are nonzero and the nonzero selection coefficients have the same sign, then there exists the polymorphism $\hat{p}$:

$$\left(\frac{b}{a+b+c}, \frac{c}{a+b+c}, \frac{a}{a+b+c}\right).$$

The crucial element of the analysis is the observation that the function

$$G(\underset{\sim}{p}) = b \ln p_1 + c \ln p_2 + a \ln p_3 \tag{2.79}$$

is constant on orbits. Indeed, (2.78) and (2.79) yield

$$\dot{G} = b\dot{p}_1 p_1^{-1} + c\dot{p}_2 p_2^{-1} + a\dot{p}_3 p_3^{-1}$$

$$= b(ap_2 - cp_3) + c(-ap_1 + bp_3) + a(cp_1 - bp_2) = 0.$$

(G is slightly more convenient than $e^G = p_1^{\,b}\,p_2^{\,c}\,p_3^{\,a}$, the natural generalization of Kimura's [1958] $p_1p_2p_3$ for $a = b = c = 1$.) In view of the symmetry of the problem, there are five distinct cases. The orbits in the p_1p_2-plane sketched in Figure 2.3 are given by $G(p_1, p_2, 1-p_1-p_2) =$ const., and the signs of $\dot{p}_i$ by (2.78).

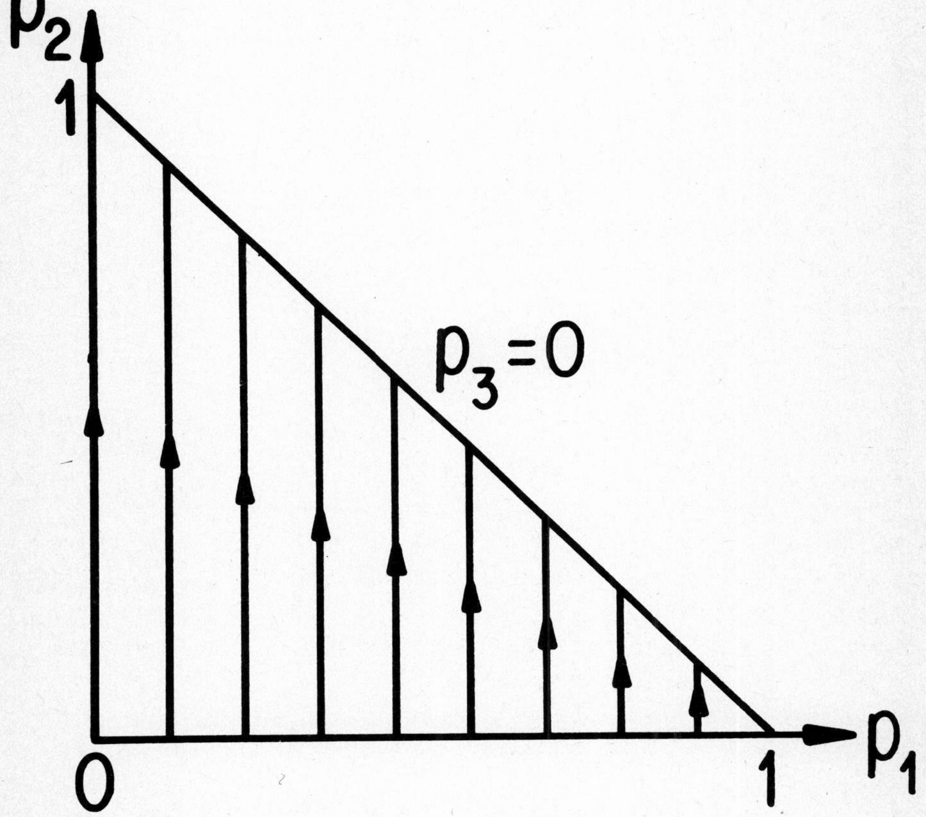

Fig. 2.3a. The triallelic competition model with $a = c = 0$, $b > 0$.

(a) $a = c = 0$, $b > 0$: A_1 does not fight and its frequency is constant. A_2 is superior to A_3, which disappears.

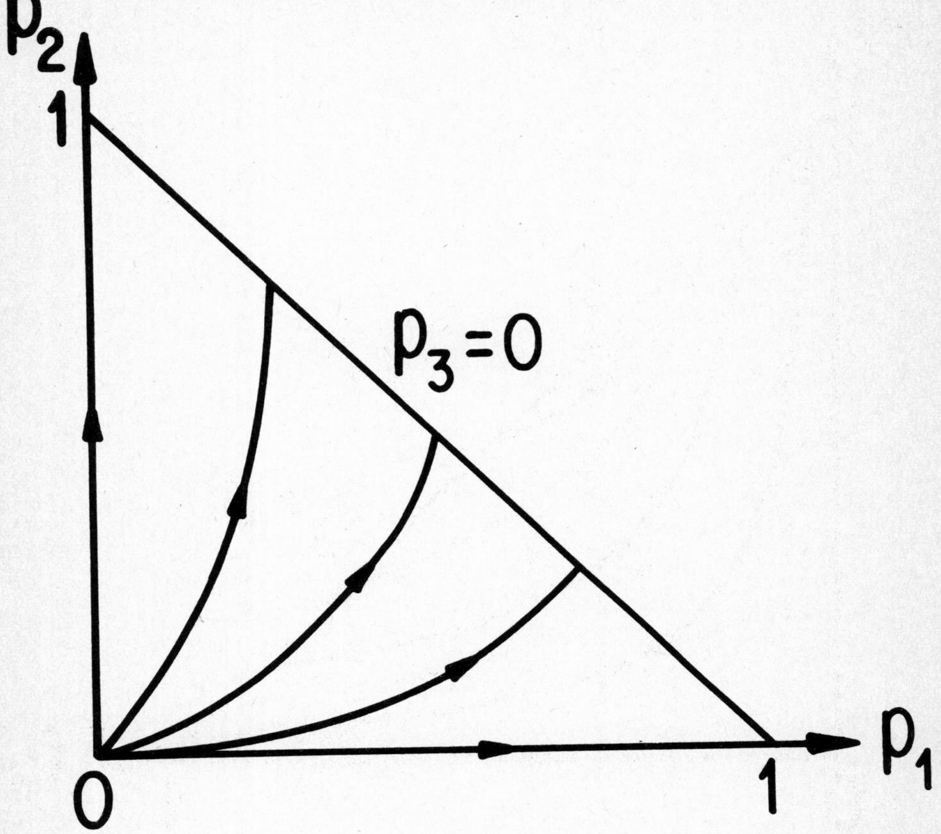

Fig. 2.3b. The triallelic competition model with $a = 0$, $b > 0$, $c < 0$.

(b) $a = 0$, $b > 0$, $c < 0$: A_3, being inferior to both A_1 and A_2, disappears. Figure 2.3b was sketched for $b > -c$. If $b = -c$, the lines emanating from the origin are straight; if $b < -c$, they curve downward.

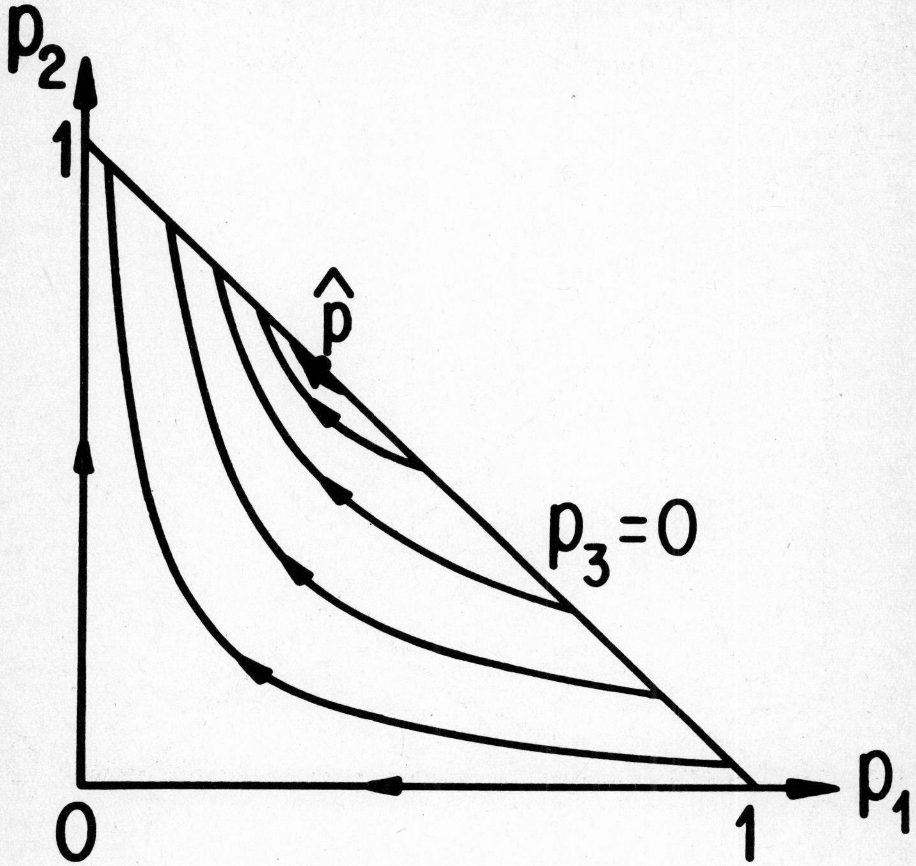

Fig. 2.3c. The triallelic competition model with $a = 0$, $b,c > 0$.

(c) $a = 0$, $b,c > 0$: The equally fit genotypes A_1 and A_2 remain in the population, while A_3 is lost.

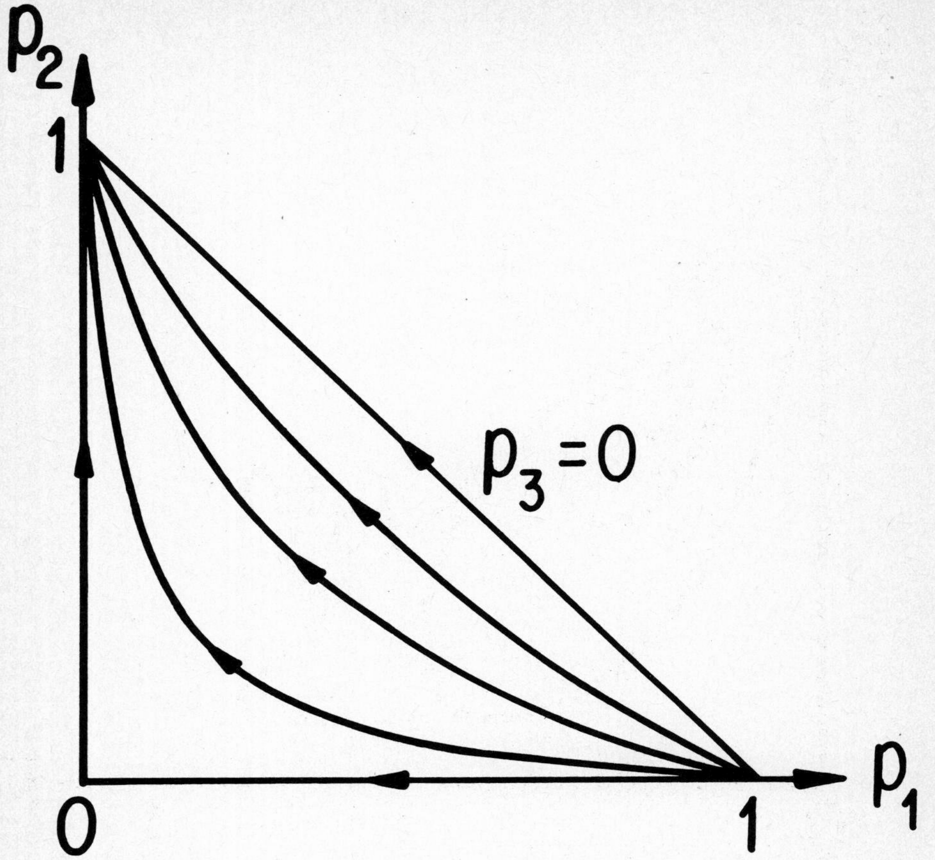

Fig. 2.3d. The triallelic competition model with $a < 0$, $b,c > 0$.

(d) $a < 0$, $b,c > 0$: A_2, which defeats both A_1 and A_3, is ultimately fixed.

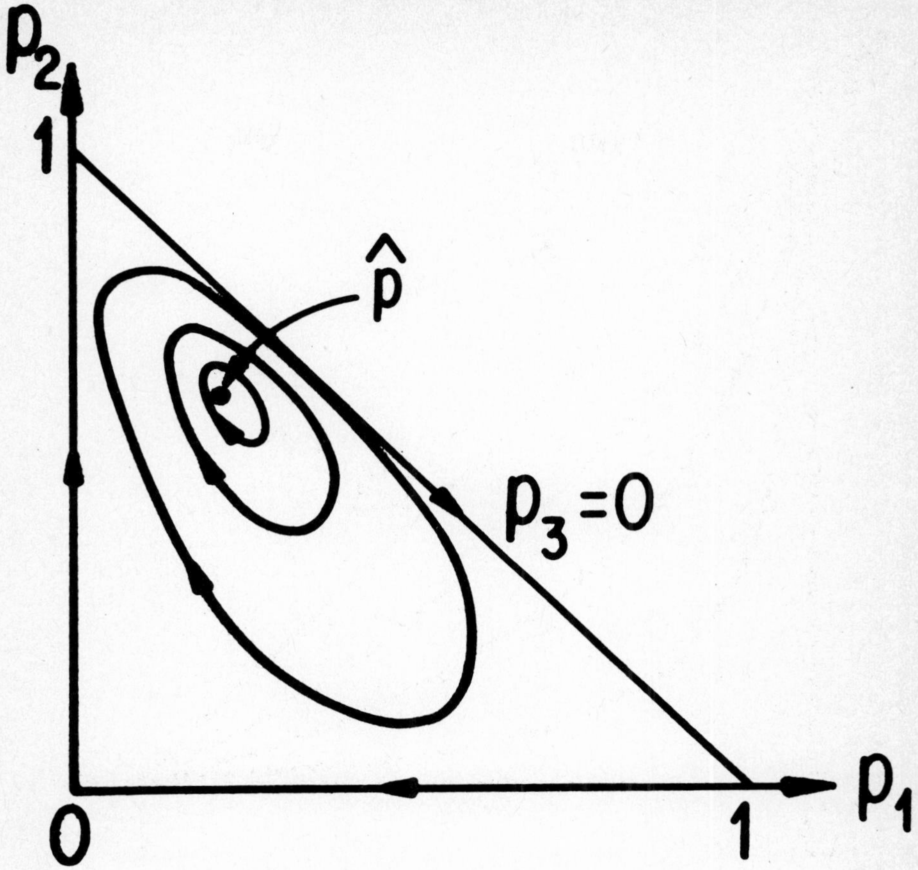

Fig. 2.3e. The triallelic competition model with $a,b,c > 0$.

(e) $a,b,c > 0$: The population cycles around $\hat{p}$. The equilibrium $\hat{p}$ is stable, but not asymptotically stable.

Before ending this section, we shall briefly consider mutation. Let u_{ij} be the probability per unit time that A_i mutates to A_j for $i \neq j$. As in Section 2.2, $u_{ii} = 0$. Instead of (2.20), we now have

$$\dot{p}_i = \sum_j p_j u_{ji} - p_i \sum_j u_{ij}. \tag{2.80}$$

To include selection, we let selection act continuously, and suppose mutation occurs every Δt time units. Separating these effects as in (2.25), from (2.63) and (2.80) we obtain

$$p_i^*(t+\Delta t) = p_i(t) + p_i(t)(m_i-\bar{m})\Delta t + O[(\Delta t)^2], \tag{2.81a}$$

$$p_i(t+\Delta t) = p_i^*(t+\Delta t) + \Big[\sum_j p_j^*(t+\Delta t)u_{ji} - p_i^*(t+\Delta t)\sum_j u_{ij}\Big]\Delta t + O[(\Delta t)^2]. \tag{2.81b}$$

Substituting (2.81a) into (2.81b) and letting $\Delta t \to 0$, we find

$$\dot{p}_i = p_i(m_i-\bar{m}) + \sum_j p_j u_{ji} - p_i \sum_j u_{ij}. \tag{2.82}$$

Thus, the approximate additivity of effects for weak selection and mutation in the discrete model of Section 2.2 becomes exact in continuous

We leave the analysis of (2.80) and (2.82) with two alleles for the problems.

In most organisms, it is necessary to distinguish zygotic and germ line genotypic frequencies, and to take into account that selection acts on the phenotype, which develops from the offspring genotype, and germ cells mutate without phenotypic effect (Nagylaki, 1974). For the usual low mutation rates, however, (2.82) is a good approximation.

2.5 *Problems*

2.1. Choose $w_1/w_2 = 1-s$, where s $(0 < s < 1)$ is a constant, in (2.18). Show that

$$t = \ln\left[\frac{q(0)p(t)}{p(0)q(t)}\right] \Big/ \ln(1-s).$$

Observe that this means that the times required to reach given frequencies for weak selection ($s << 1$) are inversely proportional to the selection intensity, and for $p(t) << p(0), q(0)$

$$t \approx -s^{-1}\ln p(t).$$

2.2. Consider two frequencies of A_1, $p = \pi_1, \pi_2$, with $\pi_1 > \pi_2$, in (2.18). If w_1 and w_2 are constant, we expect $\pi_1' > \pi_2'$. Prove this by deducing that

$$\frac{dp'}{dp} = \frac{w_1 w_2}{\bar{w}^2} > 0.$$

2.3. Consider the multiallelic discrete model with the life cycle

$$\underset{p_i}{\text{Offspring}} \xrightarrow{\text{viability}} \underset{\bar{p}_i}{\text{Adult}} \xrightarrow{\text{mutation}} \underset{\tilde{p}_i}{\text{Adult}} \xrightarrow{\text{fertility}} \underset{p_i'}{\text{Offspring}}$$

(a) Write the recursion relations.

(b) Show that if there is no differential viability ($v_i = 1$) or no differential fertility ($f_i = 1$), this system is equivalent to (2.25).

(c) Prove that if selection and mutation are weak, so that the equations may be expanded to first order, the change in the gene frequency reduces to the sum of changes due to the three factors operating, each change being calculated as if the corresponding factor were the only one present.

(d) Demonstrate that to first order the model is the same as (2.25).

2.4. Show directly from (2.41) that $0 \leq p_+ \leq 1 < p_-$, and if $u+v \leq 1$, then $\lambda \geq 0$.

2.5. Choose $m_2 - m_1 = s > 0$ in (2.71) and show that the time required to reach a specified allelic frequency $p(t)$ is

$$t = \frac{1}{s} \ln \left[\frac{p(0)q(t)}{q(0)p(t)}\right].$$

Note that this is the same as the weak-selection approximation for discrete time, and $t \approx -s^{-1} \ln p(t)$ for $p(t) << p(0), q(0)$.

2.6. For the gene frequency equation $\dot{p} = f(p)$, find a function whose derivative is positive except at equilibrium, where it is zero.

2.7. Analyze completely (2.71) with $m_i = r_i(1 + a_i p_i)^{-1}$, $a_i, r_i > 0$, $i = 1,2$.

2.8. Taking $u = u_{12}$ and $v = u_{21}$ for two alleles in (2.80) find $p(t) = p_1(t)$.

2.9. Add selection to Problem 2.8 with $m_2 - m_1 = s > 0$. Give a complete analysis. Show that if $u+v << s$, the equilibrium is the same as that for the discrete case with weak selection and mutation.

2.10. Demonstrate that the total birth rate, $\nu_i(t,0)$, satisfies the integral equation

$$\nu_i(t,0) = \int_t^{\infty} \nu_i(0,x-t)l_i(t,x)b_i(t,x)dx + \int_0^t \nu_i(t-x,0)l_i(t,x)b_i(t,x)dx.$$

Given the initial age distribution, the first integral, in principle, is known. If this integral equation is solved, then (2.47) determines $\nu_i(t,x)$ as a function of $\nu_i(0,x)$, $b_i(t,x)$, and $l_i(t,x)$.

3. PANMICTIC POPULATIONS

Hereafter we shall be concerned with diploid populations. We begin by investigating in this chapter the genetic structure of a randomly mating population in the absence of selection, mutation, and random drift. This part of population genetics was the first to be understood, and a thorough grasp of its principles is required for the formulation and interpretation of most evolutionary models. To assure the desired comprehension, some fairly detailed examples and problems, of a type that has proved useful in human genetics, are presented. In Section 3.1 we shall derive the Hardy-Weinberg law. We shall proceed to sex-linkage and two loci in Sections 3.2 and 3.3, and discuss the effect of population subdivision in Section 3.4. These sections provide background for Chapters 4 and 5, 7, 8, and 6, respectively.

The material in this chapter is based entirely on a few elementary genetic principles and the following simple rules of probability. Let $P(A)$ and $P(B)$ designate the probabilities of events A and B. We denote the probability of either A or B or both by $P(A+B)$. If $P(AB)$ represents the probability of both A and B, called the *joint* probability of A and B, then we have

$$P(A+B) = P(A)+P(B)-P(AB).$$

For *mutually exclusive* events, $P(AB) = 0$. We shall also need *conditional* probabilities: the probability of A given B is $P(A|B)$. Clearly,

$$P(AB) = P(A|B)P(B).$$

If A and B are *independent*, $P(A|B) = P(A)$, so $P(AB) = P(A)P(B)$.

3.1 *The Hardy-Weinberg Law*

We assume generations are discrete and nonoverlapping, as is true, for instance, for annual plants and many insects. If the population is *monoecious*, i.e., every individual has both male and female sexual organs, as in most plants, it can be described by one set of genotypic proportions. It will be clear that this may also be done if the initial genotypic frequencies are identical in the two sexes of a *dioecious* population. We shall show after treating the monoecious case that, if the initial genotypic frequencies are different in the two sexes, the attainment of equilibrium is merely delayed by one generation.

Denoting the alleles at the locus under consideration by A_i, the diploid individuals in the population will have genotypes A_iA_j. Mendel's Law of Segregation tells us that a gamete from an A_iA_j individual is equally likely to carry A_i and A_j. Unless otherwise stated, we shall always employ ordered genotypic frequencies. Thus, P_{ij} $(= P_{ji})$ designates the frequency of ordered A_iA_j genotypes, i.e., P_{ii} is the frequency of A_iA_i *homozygotes*, and $2P_{ij}$, $i \neq j$, is the frequency of unordered A_iA_j *heterozygotes*. The frequency of A_i in the population, p_i, reads

$$p_i = \sum_j P_{ij}. \tag{3.1}$$

By Mendel's Law of Segregation, p_i is the frequency of A_i in the gametic output of the population.

If mating occurs without regard to the genotype at the A-locus, random union of gametes yields the genotypic proportions

$$P'_{ij} = p_ip_j \tag{3.2}$$

in the next generation. Therefore, the gene frequencies do not change,

$$p'_i = \sum_j P'_{ij} = p_i, \tag{3.3}$$

and Hardy-Weinberg proportions,

$$P'_{ij} = p'_i p'_j, \tag{3.4}$$

are attained in a single generation. For two alleles, various simple numberical cases were worked out first, then Weinberg (1908) obtained the result assuming the initial absence of heterozygotes, and, simultaneously and independently, Hardy (1908) analyzed the most general situation. Weinberg (1909) extended the principle to multiple alleles.

The most important aspect of the Hardy-Weinberg law is the constancy of the allelic frequencies. This means that, if no evolutionary forces are acting, under random mating, genetic variability is conserved. The feature that equilibrium is reached in a single generation, and thereafter the genotypic frequencies are uniquely determined by the gene frequencies, is also of great significance. It allows one to analyze the evolution of a population with k alleles in terms of k-1 independent allelic frequencies, rather than $\frac{1}{2}k(k+1)-1$ independent genotypic frequencies. This is a considerable simplification. Note that the frequencies of the genotypes A_iA_j are the terms in the expansion of $(\sum_i p_i)^2$. For k alleles, we have

$$\sum_{i=1}^{k} p_i^2 \geq \frac{1}{k}.$$

Therefore, the total frequency of homozygotes is at least $1/k$, with equality if and only if $p_i = 1/k$ for all i.

The very brief derivation of (3.4) presented above is directly applicable if a large number of sperm and eggs are produced, and some of these fuse at random to form zygotes, as exemplified by some marine organisms. For most species, however, considering matings explicitly approximates the biological situation more closely. This approach has the advantage of generalizing readily to nonrandom mating and differential fertility.

Let the frequency of ordered $A_iA_j \times A_kA_l$ matings be $X_{ij,kl}$ (e.g., the total proportion of $A_1A_2 \times A_1A_1$ matings, without regard to allelic and genotypic order, is $4X_{12,11}$). We shall impose the restriction of panmixia only when necessary. The mating frequencies satisfy the symmetry conditions $X_{ij,kl} = X_{ji,kl} = X_{kl,ij}$ and the normalization

$$\sum_{ijkl} X_{ij,kl} = 1. \tag{3.5}$$

To calculate the genotypic proportions in the next generation, we distinguish homozygotes from heterozygotes throughout. We must include, due to ordering, a factor of 2 for all heterozygotes and for all matings between distinct genotypes, and, due to segregation, one of $\frac{1}{2}$ for all heterozygotes. In addition, we need to observe that the cross $A_iA_j \times A_iA_j$, $i \neq j$, can produce A_iA_j in 2 ways. Hence, we obtain

$$P'_{ii} = X_{ii,ii} + \sum_{k\neq i} 2(2)\tfrac{1}{2}X_{ik,ii} + \sum_{k\neq i} 2(2)(\tfrac{1}{2})\tfrac{1}{2}X_{ik,ki}$$
$$+ \sum_{k\neq i} \sum_{l\neq i,\, l>k} 2(2)(2)(\tfrac{1}{2})\tfrac{1}{2}X_{ik,li},$$

$$2P'_{ij} = 2X_{ii,jj} + \sum_{k\neq i} 2(2)\tfrac{1}{2}X_{ik,jj} + \sum_{l\neq j} 2(2)\tfrac{1}{2}X_{ii,lj}$$
$$+ 2(2)(2)(\tfrac{1}{2})\tfrac{1}{2}X_{ij,ij} + \sum_{k\neq i} \sum_{\substack{l\neq j \\ (k,l)\neq(j,i)}} 2(2)(2)(\tfrac{1}{2})\tfrac{1}{2}X_{ik,lj}, \qquad i \neq j.$$

Removing the restrictions from the sums and combining terms yields

$$P'_{ij} = \sum_{kl} X_{ik,lj}, \qquad \text{for all } i,j. \tag{3.6}$$

At this point, we posit random mating:

$$X_{ik,lj} = P_{ik}P_{lj}. \tag{3.7}$$

Observe that (3.7) satisfies (3.5). Substituting (3.7) into (3.6) immediately gives (3.2).

We proceed to generalize to separate sexes with different initial genotypic frequencies. Let P_{ij} and Q_{ij} be the ordered frequencies of A_iA_j in males and females. The gene frequencies in the two sexes are

$$p_i = \sum_j P_{ij}, \qquad q_i = \sum_j Q_{ij}. \tag{3.8}$$

With random union of gametes, we have the same genotypic frequencies in the two sexes in the next generation:

$$P'_{ij} = Q'_{ij} = \tfrac{1}{2}(p_iq_j + p_jq_i). \tag{3.9}$$

Therefore, the allelic frequencies in each sex are just the average allelic frequencies in the previous generation:

$$p'_i = q'_i = \tfrac{1}{2}(p_i+q_i). \tag{3.10}$$

Thus, another generation of random union of gametes produces Hardy-Weinberg proportions

$$P''_{ij} = Q''_{ij} = p'_ip'_j. \tag{3.11}$$

To derive (3.9) from the mating structure, we now define $X_{ij,kl}$ as the ordered frequency of matings between A_iA_j males and A_kA_l females. We still have $X_{ij,kl} = X_{ji,kl} = X_{ij,lk}$, but reciprocal crosses are not necessarily equally frequent: $X_{ij,kl} \neq X_{kl,ij}$. Equation (3.5) still applies, but now we must generalize (3.6) to

$$P_{ij} = Q_{ij} = \tfrac{1}{2}\sum_{kl}(X_{ik,lj} + X_{lj,ik}), \tag{3.12}$$

and random mating now means

$$X_{ij,kl} = P_{ij}Q_{kl}. \tag{3.13}$$

Note that (3.13) is consistent with (3.5). Inserting (3.13) into (3.12) and recalling (3.8) at once produces (3.9).

It will be convenient to refer to (3.9), and to the corresponding relation for a sex-linked locus, as generalized Hardy-Weinberg proportions.

We end this section with some examples.

1. The *MN* blood group:

There are two alleles, M and N, the three genotypes being easily distinguishable. This is the case of *no dominance*. The frequency, p_1, of M can be obtained from the measured genotypic proportions, P_{11}, $2P_{12}$, P_{22}, of MM, MN, NN using $p_1 = P_{11}+P_{12}$, and Hardy-Weinberg can be tested with one degree of freedom (see, e.g., Li, 1955; Crow and Kimura, 1970).

2. Phenylthiocarbamide (PTC):

The genotypes A_1A_1 and A_1A_2 can taste PTC strongly, while A_2A_2 individuals either cannot or can do so only very faintly. In general, if A_1A_2 has the same *phenotype* (i.e., expressed character) as A_1A_1, we say A_1 is *dominant* (to A_2) and A_2 is *recessive* (to A_1). The frequency, q, of A_2 is given by $q = \sqrt{P_{22}}$, but the genetic mechanism for a dominant gene cannot be tested without using some family data. If homozygous recessives are affected by some disease, e.g., phenylketonuria (PKU), then usually $q \ll 1$. This has the important implication that the frequency of carriers of the recessive allele, $2pq \approx 2q$, is much greater than the proportion affected, q^2. In applications to diseases, care must be exercised; owing to selection, the recessive frequency will usually be less in adults than in zygotes. Since A_2A_2 is extremely rare, however, the error is generally quite small.

3. The *ABO* blood group:

For our purposes, it will suffice to consider three alleles, A, B, O, with A and B dominant to O (for more detail, see Cavalli-Sforza and Bodmer, 1971), as shown in Table 3.1.

TABLE 3.1

The ABO *blood group*

Blood Type	Genotypes	Frequency
A	AA, AO	$p_A^2 + 2p_Ap_O$
B	BB, BO	$p_B^2 + 2p_Bp_O$
AB	AB	$2p_Ap_B$
O	OO	p_O^2

If R_A, R_B, R_{AB}, and R_O are the observed frequencies of the blood types, we have $p_O = \sqrt{R_O}$ and $(p_A+p_O)^2 = R_A+R_O$, whence $p_A = \sqrt{R_A+R_O} - \sqrt{R_O}$. As for two alleles without dominance, we can test Hardy-Weinberg with one degree of freedom (Li, 1955; Crow and Kimura, 1970).

4. The parentage of homozygous recessives:

Here and in the following two examples, let A be dominant to the other allele, a. Homozygous recessives all come from the matings $AA \times AA$, $AA \times Aa$, $aa \times aa$. We wish to compute the proportion of aa individuals originating in each of the three matings. The calculation set out in Table 3.2 shows that if the recessive is rare, $q << 1$, almost all affected individuals come from unions of two carriers. Selection against homozygous recessives will again produce a small error here.

TABLE 3.2

The parentage of homozygous recessives

Mating	Freq. of Mating	Freq. of aa Progeny	Freq. Among aa Progeny
$Aa \times Aa$	$(2pq)^2$	$\frac{1}{4}(4p^2q^2)$	p^2
$Aa \times aa$	$2(2pq)q^2$	$\frac{1}{2}(4pq^3)$	$2pq$
$aa \times aa$	$(q^2)^2$	q^4	q^2

5. Parent-offspring relations:

It will often be convenient to abbreviate the dominant and recessive phenotypes by D and R.

(a) Breeding ratios: These are the conditional probabilities of the birth of a child of phenotype i (= D,R) in a $j \times k$ mating, denoted by $P_{i,jk}$. Obviously, $P_{R,RR} = 1$. We need not resort to a mating table to calculate Snyder's (1932) ratios, $P_{R,DR}$ and $P_{R,DD}$. When applicable, analysis in terms of gametes is much quicker. Let x be the probability of drawing a gamete carrying a from a dominant phenotype. Then $P_{R,DR} = x$ and $P_{R,DD} = x^2$, where

$$x = \frac{(1/2)2pq}{p^2 + 2pq} = \frac{q}{1+q}. \tag{3.14}$$

As a check, note that recessives have been correctly allocated to the three types of matings:

$$(q^2)^2 P_{R,RR} + 2q^2(1-q^2)P_{R,DR} + (1-q^2)^2 P_{R,DD} = q^2.$$

Parent-offspring resemblance is revealed by the inequalities $P_{R,RR} > P_{R,DR} > P_{R,DD}$. The conditional probabilities of dominants are given by $P_{D,ij} = 1-P_{R,ij}$. Family frequencies are calculated by multiplying the conditional probabilities by the appropriate mating frequencies.

(b) Mother-child probabilities: We desire to compute the probability that a mother (or father) of phenotype j has a child of phenotype i, $P_{i,j}$. This can be done, of course, from the breeding ratios:

$$P_{i,j} = q^2 P_{i,jR} + (1-q^2) P_{i,jD}.$$

But an even faster method is to note that a randomly chosen gamete carries A and a with probabilities p and q, so, with panmixia, $P_{R,R} = q$ and $P_{R,D} = xq$. Mother and child resemble each other, for $P_{R,R} > P_{R,D}$. As a check, observe here that $q^2 P_{R,R} + (1-q^2)P_{R,D} = q^2$.

6. A sibling distribution:

We wish to derive the probability, F_{sn}, that a sibship of size s has exactly n (= 0,1,...,s) dominant individuals (Cotterman, 1937). We must separate the six genotypically distinct matings, as done in Table 3.3.

TABLE 3.3

The sibling distribution for a dominant gene

Mating	Freq. of Mating	Conditional Probability of Sibship		
		$n = s$	$0 < n < s$	$n = 0$
$AA \times AA$	$(p^2)^2$	1	0	0
$AA \times Aa$	$2p^2(2pq)$	1	0	0
$AA \times aa$	$2p^2q^2$	1	0	0
$Aa \times Aa$	$(2pq)^2$	$(\frac{3}{4})^s$	$\binom{s}{n}(\frac{3}{4})^n(\frac{1}{4})^{s-n}$	$(\frac{1}{4})^s$
$Aa \times aa$	$2(2pq)q^2$	$(\frac{1}{2})^s$	$\binom{s}{n}(\frac{1}{2})^n(\frac{1}{2})^{s-n}$	$(\frac{1}{2})^s$
$aa \times aa$	$(q^2)^2$	0	0	1

We obtain $F_{sn}(q)$ from the sums of the unconditional probabilities in the columns of Table 3.3. With some simplifications, we find

$$F_{ss} = p\,\{\,p(2-p^2) + q^2\left[4(\tfrac{3}{4})^s p + (\tfrac{1}{2})^{s-2} q\right]\},\qquad (3.15a)$$

$$F_{s0} = q^2\left[\left(\tfrac{1}{2}\right)^{s-1}p + q\right]^2, \tag{3.15b}$$

$$F_{sn} = \left(\tfrac{1}{2}\right)^{s-2}\binom{s}{n}pq^2\left[3^n\left(\tfrac{1}{2}\right)^s p + q\right], \qquad 0 < n < s. \tag{3.15c}$$

As partial checks, we observe that (3.15) gives correctly the trivial results $F_{11} = 1-q^2$ and $F_{10} = q^2$, and with a little algebra we may verify that

$$\sum_{n=0}^{s} F_{sn} = 1. \tag{3.16}$$

For two siblings, (3.15) yields

$$F_{22} = p[1+pq(1+\tfrac{1}{4}q)],$$

$$F_{20} = \tfrac{1}{4}q^2(1+q)^2,$$

$$F_{21} = \tfrac{1}{2}pq^2(3+q).$$

The probability that there are precisely n dominants among s randomly chosen individuals is

$$f_{sn} = \binom{s}{n}(1-q^2)^n(q^2)^{s-n},$$

whence

$$f_{22} = p^2(1+q)^2, \qquad f_{20} = q^4, \qquad f_{21} = 2pq^2(1+q).$$

It is easy to see that concordant sib pairs are more frequent and discordant ones less than concordant and discordant pairs among randomly chosen individuals:

$$F_{22} > f_{22}, \qquad F_{20} > f_{20}, \qquad F_{21} < f_{21}. \tag{3.17}$$

3.2 X-*Linkage*

In most higher animals, such as mammals, sex is determined by a single chromosomal difference between males and females. One sex has a pair of *sex chromosomes* of the same type, *XX*, and the other has two different ones, denoted *XY*. Although in some taxa (e.g., birds) *XX* is male and *XY* is female, we shall always refer to *XX* as female and *XY* as male, this situation being far more common (e.g., mammals and *Drosophila*). Chromosomes that are not sex chromosomes are called *autosomes*, and the genes on them are *autosomal*. Genes on the sex chromosomes are said to be *sex-linked*; the terms *X-linked* and *Y-linked* distinguish genes on the *X* and *Y* chromosomes. Since there are much fewer Y-linked loci with known phenotypic effects than X-linked ones, sex- and X-linkage are often used synonymously. It is apparent that, if we interpret all variables as referring only to males, Chapter 2 applies to the dynamics of Y-linked genes. Since X-linked recessives are expressed in males, X-linked loci are of particular interest in human and *Drosophila* genetics.

Let the frequency of A_i be p_i in males and q_i in females. We denote the proportion of A_iA_j females by Q_{ij}. Random union of gametes informs us that

$$p_i' = q_i, \tag{3.18a}$$

$$Q_{ij}' = \frac{1}{2}(p_iq_j + p_jq_i). \tag{3.18b}$$

To deduce (3.18) from explicit consideration of matings, let $X_{i,jk}$ designate the ordered frequency of unions between A_i males and A_jA_k females. Thus,

$$\sum_{ijk} X_{i,jk} = 1. \tag{3.19}$$

Then

$$p_i' = \sum_{jk} X_{j,ik}, \tag{3.20a}$$

$$Q_{ij}' = \frac{1}{2}\sum_k (X_{i,kj} + X_{j,ik}). \tag{3.20b}$$

With random mating,

$$X_{i,jk} = p_iQ_{jk}, \tag{3.21}$$

noting that

$$q_i = \sum_j Q_{ij}, \tag{3.22}$$

(3.20) reduces to (3.18).

From (3.18b) we find the expected relation

$$q_i' = \frac{1}{2}(p_i + q_i). \tag{3.23}$$

Using (3.18a) and (3.23), we see that the frequency of A_i in the entire population, $x_i = (1/3)(p_i + 2q_i)$, is conserved: $x_i' = x_i$, whence $x_i(t) = x_i(0)$. The male-female gene frequency difference, $y_i = p_i - q_i$, changes sign and is reduced by 1/2 every generation: $y_i' = -(1/2)y_i$, whence $y_i(t) = y_i(0)(-1/2)^t$. Therefore,

$$p_i(t) = x_i(0) + \frac{2}{3}(-\frac{1}{2})^t y_i(0), \tag{3.24a}$$

$$q_i(t) = x_i(0) - \frac{1}{3}(-\frac{1}{2})^t y_i(0). \tag{3.24b}$$

Thus, the Hardy-Weinberg equilibrium

$$p_i = x_i(0), \qquad Q_{ij} = x_i(0)x_j(0), \tag{3.25}$$

is approached in a rapid, oscillatory manner, but, in contrast to the autosomal dioecious case, it is not attained in two generations.

For two alleles, numerical examples of the behavior (3.24) were first given by Jennings (1916); the analytic solution was deduced by Robbins (1918).

An important consequence of (3.25) is that, if q is the frequency of a rare recessive allele, the corresponding trait (various types of color blindness are examples) is expressed with a much higher frequency, q, in males than in females, in whom the frequency is q^2.

As an example, suppose A, with frequency p, is completely dominant to a, and the population is in equilibrium. What are the parent-offspring probabilities, $P_{i,j}$, of Example 5(b) of Section 3.1? We shall use an asterisk to refer to females.

For father-son, clearly, $P_{R,R} = P_{R,D} = q$, indicating that the father's genotype is irrelevant since he contributes a Y-chromosome to his son. Father-daughter is equally simple: $P_{R*,R} = q$, $P_{R*,D} = 0$. For mother-son, we evidently have $P_{R,R*} = 1$, and, since $x = q/(1+q)$ is the probability of drawing a from a dominant mother, we see that $P_{R,D*} = x$. The mother-daughter relation is the same as for an autosome: $P_{R*,R*} = q$, $P_{R*,D*} = xq$. The following relations are satisfied, as they must be:

$$qP_{R,R} + pP_{R,D} = q,$$

$$qP_{R*,R} + pP_{R*,D} = q^2,$$

$$q^2P_{R,R*} + (1-q^2)P_{R,D*} = q,$$

$$q^2P_{R*,R*} + (1-q^2)P_{R*,D*} = q^2.$$

3.3 *Two loci*

We shall suppose that the sexes need not be distinguished; the dioecious case is treated in Problem 3.10. Let the alleles at the A- and B-loci, A_i and B_i, have frequencies p_i and q_i, the number of alleles at each locus being arbitrary. We denote the frequency of A_iB_j gametes in the gametic output of generation t by $P_{ij}(t)$. The gametic frequencies are our basic variables; the gene frequencies

$$p_i = \sum_j P_{ij}, \qquad q_j = \sum_i P_{ij}, \tag{3.26}$$

no longer suffice to describe the population.

Gametes A_iB_j and A_kB_l unite to form an individual A_iB_j/A_kB_l. A proportion $1-c$ of the gametes produced by this individual are parental (or nonrecombinant) gametes, i.e., $\frac{1}{2}(1-c)$ are A_iB_j and $\frac{1}{2}(1-c)$ are A_kB_l, and a fraction c are nonparental (or recombinant), i.e., $\frac{1}{2}c$ A_iB_l and $\frac{1}{2}c$ A_kB_j. If the two loci are on the same chromosome, the *recombination* (or *cross-over*) frequency c satisfies $0 \leq c \leq 1/2$. The numerical value of c depends on the structure of the chromosome and the position of the two loci. If we measure distance y along the chromosome from the A-locus in a particular direction, the recombination fraction between A and a locus at y is a monotone increasing function of y. We shall assume $c > 0$ because it is manifest that, for our purposes, $c = 0$ corresponds to a single locus. For loci on different chromosomes, almost invariably, $c = 1/2$, as expected. Loci with $c = 1/2$ are called *independent* (or *unlinked*).

A proportion $1-c$ of the gametes in generation $t+1$ are produced without recombination. Clearly, the contribution to P'_{ij} from these events is $(1-c)P_{ij}$. Assuming random mating, the contribution of recombinant events is cp_iq_j. Therefore,

$$P'_{ij} = (1-c)P_{ij} + cp_iq_j. \tag{3.27}$$

It follows immediately from (3.26) and (3.27) that the gene frequencies are conserved, as they must be. We define the *linkage disequilibria*,

$$D_{ij} = P_{ij} - p_iq_j, \tag{3.28}$$

in order to have measures of the departure from random combination of alleles within gametes. From (3.27) and (3.28) we deduce at once

$$D_{ij} = (1-c)D_{ij}, \tag{3.29}$$

whence

$$D_{ij}(t) = D_{ij}(0)(1-c)^t. \tag{3.30}$$

We conclude that $D_{ij} \to 0$ at the rate $1-c$. Thus, *linkage equilibrium* (or *gametic phase equilibrium*), $P_{ij} = p_iq_j$, is approached gradually without oscillation. The larger c, the faster is the rate of convergence, the most rapid being $(1/2)^t$ for independent loci.

For two alleles at each locus, numerical cases were worked out by Jennings (1917); the analytic solution was obtained by Robbins (1918a). Jennings and Robbins considered the gametic output of all the genotypes, as we shall do in Chapter 8; the elegant argument used here to deduce (3.27) is due to Malécot (1948). Geiringer (1944) showed that even with multiple loci and multiple alleles the gametic frequencies converge to products of the appropriate allelic frequencies. Bennett (1954) gave a simplified proof.

The ordered frequency of A_iB_j/A_kB_l is $P_{ij}P_{kl}$, provided there has been at least one generation of random mating. By summing $P_{ij}P_{kl}$ over j,l and i,k, we recover the Hardy-Weinberg law at each locus. In linkage equilibrium, $P_{ij}P_{kl}$ simplifies to $(p_ip_k)(q_jq_l)$, the products of the single-locus Hardy-Weinberg genotypic frequencies. This observation has the important consequence that, since linkage has no effect on equilibrium genotypic frequencies, family data are required to detect linkage. Note also that in linkage equilibrium the *coupling* (A_iB_j/A_kB_l) and *repulsion* (A_iB_l/A_kB_j) phases are equally frequent. We may combine (3.27) and (3.28) to obtain

$$\Delta P_{ij} = -cD_{ij}. \qquad (3.31)$$

For two alleles at each locus, the usual notation is $x_1 = P_{11}$, $x_2 = P_{12}$, $x_3 = P_{21}$, $x_4 = P_{22}$. One finds easily from (3.28)

$$D_{11} = D_{22} = -D_{12} = -D_{21} = x_1x_4 - x_2x_3 \equiv D, \qquad (3.32)$$

so that (3.31) becomes

$$\Delta x_i = -\varepsilon_i cD, \qquad (3.33)$$

where $\varepsilon_1 = \varepsilon_4 = 1$ and $\varepsilon_2 = \varepsilon_3 = -1$.

As an example, let us consider *duplicate genes*. For any number of loci, let the alleles at locus i be $A^{(i)}$ and $a^{(i)}$, with frequencies $p^{(i)}$ and $q^{(i)}$. An individual has phenotype R (recessive) if he is homozygous for $a^{(i)}$ at every locus; otherwise his phenotype is D (dominant).

(a) Breeding ratios: We designate the probability of drawing an R gamete, i.e., one with $a^{(i)}$ at every locus, from a D individual by y. Evidently, $P_{R,RR} = 1$, $P_{R,DR} = y$, $P_{R,DD} = y^2$, as for a single locus. If the population is in linkage equilibrium, for any recombination frequencies, the single-locus result generalizes to $y = z/(1+z)$, where $z = \prod_i q^{(i)}$ is the proportion of R gametes. To see this explicitly for two diallelic loci, suppose the alleles $A,a;B,b$ have frequencies $p,q;u,v$. Then the gametic output of the genotypes *AaBb*, *Aabb*, *aaBb* yields

$$y = \frac{(1/4)4pquv + (1/2)2pqv^2 + (1/2)2q^2uv}{1 - q^2v^2} = \frac{qv}{1+qv}.$$

(b) Sibling distribution: We shall calculate in the diallelic two-locus case the probability, Q_{sn}, that there are exactly n D individuals among s siblings. Although in linkage equilibrium the two types of double heterozygotes, AB/ab and Ab/aB, are equally frequent in the population, within a family, one must take into account that these have different gametic outputs, which depend on c. The extreme case $c = 0$ is trivial: $Q_{sn} = F_{sn}(qv)$, F_{sn} being the single-locus distribution given (3.15).

Let us derive Q_{sn} for independent loci, $c = 1/2$. It is convenient to use d and r for phenotypes at one locus; thus, dr denotes an individual with the dominant phenotype at the A-locus (i.e., AA or Aa) and the recessive at the other (i.e., bb). Let the number of dd, dr, rd, rr siblings be i, j, $n-i-j$, $s-n$. For independent loci in linkage equilibrium, the probability of having $(i+j)$ d- sibs and $(n-j)$ $-d$ ones is $F_{s,i+j}(q)F_{s,n-j}(v)$. Given that we have $(i+j)$ d- sibs and $(n-j)$ $-d$ ones, we require the probability of choosing i dd sibs from $(i+j)$ d-ones and $(n-i-j)$ rd sibs from $(s-i-j)$ r- ones. Since we have conditioned on choosing $(n-j)$ $-d$ individuals, the desired probability is

$$\binom{i+j}{i}\binom{s-i-j}{n-i-j} \Big/ \binom{s}{n-j}.$$

Hence,

$$Q_{sn} = \sum_{i=0}^{n} \sum_{j=0}^{n-i} \left[\binom{i+j}{i}\binom{s-i-j}{s-n} \Big/ \binom{s}{n-j}\right] F_{s,i+j}(q)F_{s,n-j}(v)$$

$$= \sum_{k=0}^{n} \sum_{l=n-k}^{n} \left[\binom{k}{n-l}\binom{s-k}{s-n} \Big/ \binom{s}{l}\right] F_{sk}(q)F_{sl}(v), \qquad (3.34)$$

where we changed variables to $k = i+j$ and $l = n-j$.

It is easy to check that (3.34) reduces sensibly for $s = 1,2$. Let us verify that

$$\sum_{n=0}^{s} Q_{sn} = 1.$$

We put $m = n-l$ to deduce

$$\sum_{n=0}^{s} Q_{sn} = \sum_{k=0}^{s} F_{sk}(q) \sum_{l=0}^{s} \left[F_{sl}(v) \Big/ \binom{s}{l}\right] \sum_{m=m_1}^{m_2} \binom{k}{m}\binom{s-k}{s-l-m},$$

where $m_1 = \max(0,k-l)$ and $m_2 = \min(k,s-l)$. But, as can be proved by writing

$$(1+x)^i(1+x)^j = (1+x)^{i+j},$$

expanding, and equating coefficients,

$$\sum_j \binom{i}{j}\binom{k-i}{l-j} = \binom{k}{l}.$$

Therefore, recalling (3.16),

$$\sum_{n=0}^{s} Q_{sn} = \sum_{k=0}^{s} F_{sk}(q) \sum_{l=0}^{s} F_{sl}(v) = 1.$$

3.4 *Population Subdivision*

Many natural populations are divided into subpopulations. We shall show that, if each subpopulation is panmictic, the effect of the subdivision is to raise the *homozygosity*, f, compared to the homozygosity of a randomly mating population with the same allelic frequencies, f_r. Let c_j and $p_i^{(j)}$ represent the proportion of individuals and the frequency of A_i in subpopulation j. Using E (expectation) to designate averages, we may write the allelic frequencies in the population as

$$\overline{p_i} \equiv E(p_i) = \sum_j c_j p_i^{(j)}.$$

The variance of p_i in the population is

$$V(p_i) = [(p_i-\overline{p_i})^2] = \overline{p_i^2} - \overline{p_i}^2. \tag{3.35}$$

Therefore, the frequency of A_iA_i in the population,

$$\overline{p_i^2} = \overline{p_i}^2 + V(p_i), \tag{3.36}$$

is higher than in a randomly mating population by the variance of p_i, which is zero only if the allelic frequencies in all the subpopulations are the same. Equation (3.36) is Wahlund's principle (Wahlund, 1928).

The proportion of homozygotes in the population reads

$$f = \sum_i \overline{p_i^2} = f_r + \sum_i V(p_i), \tag{3.37}$$

with

$$f_r = \sum_i \overline{p_i}^2.$$

The decrease in *heterozygosity*, h, is given by

$$h = 1-f = h_r - \sum_i V(p_i), \tag{3.38}$$

where $h_r = 1-f_r$, the proportion of heterozygotes with random mating. For two alleles, observing that $V(p) = V(q) \equiv V$, (3.36) and (3.38) yield

$$P_{11} = \bar{p}^2 + V, \qquad P_{12} = \bar{p}\bar{q} - V, \qquad P_{22} = \bar{q}^2 + V.$$

3.5 *Problems*

3.1. The genotypic frequencies at a diallelic locus in a monoecious population can be represented on a de Finetti (1926) diagram, sketched in Figure 3.1. With $D = P_{11}$, $H = 2P_{12}$, $R = P_{22}$, $D+H+R = 1$, so the distances from a point to the sides of an equilateral triangle of unit altitude can be taken to correspond to D, H, and R. Prove that the Hardy-Weinberg curve, $H^2 = 4DR$, is a parabola, and the vertical line divides the base of the triangle in the ratio of $p{:}q$.

3.2. Consider an arbitrary monoecious diallelic population. Using $Q = P_{12} - p_1p_2$ as abscissa and $p = p_1$ as ordinate, find the equations of the curves bounding the permitted region in the Qp-plane, and sketch this region. The fact that Hardy-Weinberg populations fall on the p-axis makes this plot convenient for displaying the evolution of populations with overlapping generations.

3.3. For a diallelic Hardy-Weinberg population, find the probabilities $P_{i,j}$ that a mother of genotype j (i,j = *AA*, *Aa*, *aa*) has a child of genotype i.

3.4. Prove (3.17).

3.5. Calculate the probability that, in a diallelic Hardy-Weinberg population, a sibship of size s has i *AA*, j *Aa*, and k *aa* individuals.

3.6. *Penetrance* is the extent to which a genotype is expressed as a phenotype. The first model considered for the inheritance of human handedness was autosomal diallelic, right being completely dominant to left, i.e., *AA* and *Aa* had phenotype R, while *aa* was L. This was immediately seen to fail because about half the progeny of $L \times L$ matings are R. A more sophisticated model

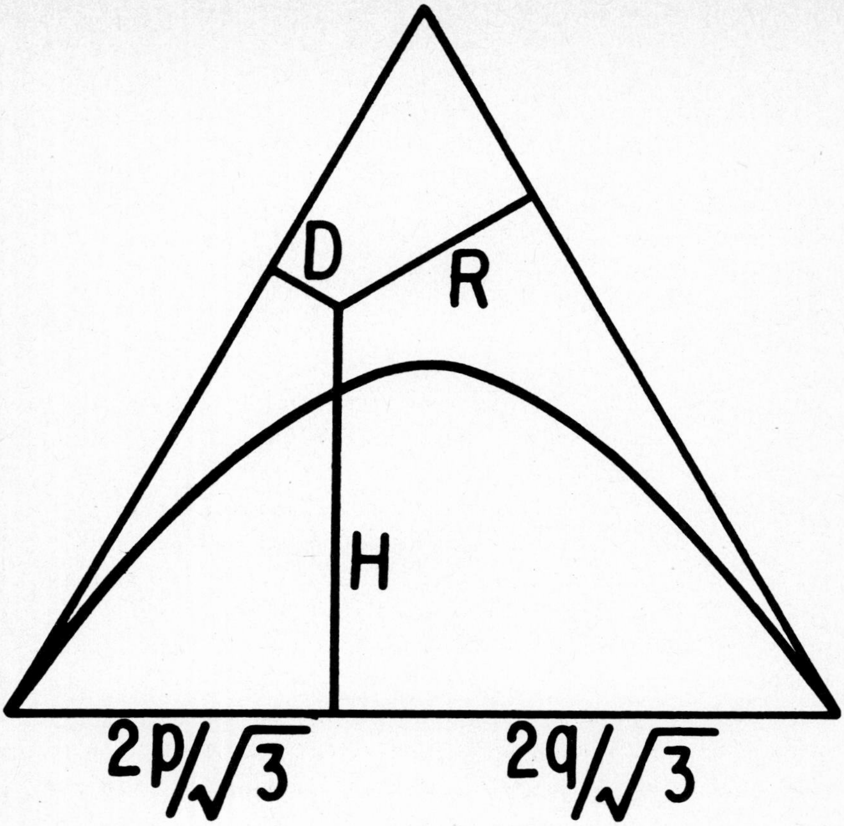

Fig. 3.1. The de Finetti diagram.

(Trankell, 1955), which still cannot account for the related phenotypic variation in cerebral organization (Levy and Nagylaki, 1972), postulates incomplete penetrance for a. Suppose AA and Aa are always R, but of aa genotypes, only a fraction x among parents and y among children are L. Assume mating is random, and A has frequency p in both generations. Notice that $x > y$ ($x < y$) implies a higher (lower) frequency of sinistrality among parents than children. For instance, $x < y$ might reflect a decrease in cultural pressure toward dextrality. Obtain the proportion of sinistral progeny in the three kinds of matings and the probability that there are exactly n sinistrals among s

siblings.

3.7. For a diallelic X-linked trait with complete dominance, assuming equilibrium, derive the probability that in a sibship precisely i of m males and j of n females are dominant.

3.8. Two populations $AABB$ and $aabb$ are crossed. If the A and B loci are independent, show that they are in linkage equilibrium after two generations. Why is (3.30) inapplicable?

3.9. A recent model for the inheritance of human handedness (Levy and Nagylaki, 1972; Nagylaki and Levy, 1973; Levy, 1976; Levy and Reid, 1976, 1977) posits the existence of two diallelic loci, one determining cerebral dominance and the other governing hand control. At the first locus, the allele L (frequency p) for left hemisphere language is completely dominant to the allele l (freq. q) for right hemisphere language. Thus, verbal and analytic functions reside primarily in the left hemisphere of LL and Ll, but in the right hemisphere of ll individuals. In each case, the other hemisphere performs mainly spatial and synthetic functions. At the second locus, the allele C (freq. u) for contralateral hand control is completely dominant to c (freq. v), which specifies ipsilateral hand control. Therefore, $LLCC$, $LLCc$, $LlCC$, $LlCc$, and $llcc$ are dextral (freq. γ), $LLcc$ and $Llcc$ are sinistral (freq. $\lambda = 1-\gamma$) with ipsilateral hand control (freq. β), and $llCC$ and $llCc$ are sinistral with contralateral hand control (freq. $\alpha = \lambda-\beta$). Panmixia and linkage equilibrium are assumed.

(a) Calculate the allelic frequencies in terms of α and β. To obtain a unique solution, use the fact that the proportion of dextrals with right hemisphere language is very small, i.e., $q^2v^2 << 1$.

(b) Derive the proportion of sinistrals in the three types of matings by using the single-locus breeding ratios.

(c) Solve (b) directly by calculating the gametic output of the two phenotypes.

(d) For unlinked loci, prove that the probability of having exactly n sinistrals among s sibs reads

$$P_{sn} = \sum_{j=0}^{n} \sum_{i=0}^{s-n} A_{snij} F_{s,s-n-i+j}(q) F_{s,s-i-j}(v),$$

where

$$A_{snij} = \binom{s-n-i+j}{j}\binom{n+i-j}{i} \Big/ \binom{s}{i+j}.$$

(e) Demonstrate that

$$\sum_{n=0}^{s} P_{sn} = 1.$$

3.10. To generalize Section 3.3 to separate sexes, suppose P_{ij} and $\tilde{P}_{ij}$, p_i and $\tilde{p}_i$, q_j and $\tilde{q}_j$, and c and $\tilde{c}$ are the frequencies of A_iB_j, A_i, B_j, and the recombination fractions in males and females.

(a) Show that with random mating the recursion relations are

$$P'_{ij} = (1-c)\tfrac{1}{2}(P_{ij}+\tilde{P}_{ij}) + c\tfrac{1}{2}(p_i\tilde{q}_j + \tilde{p}_iq_j),$$

$$\tilde{P}'_{ij} = (1-\tilde{c})\tfrac{1}{2}(P_{ij}+\tilde{P}_{ij}) + \tilde{c}\tfrac{1}{2}(p_i\tilde{q}_j + \tilde{p}_iq_j).$$

(b) Deduce that, after one generation, the gene frequencies in the two sexes are constant and equal to each other, being given by

$$\overline{p}_i = \tfrac{1}{2}[p_i(0) + \tilde{p}_i(0)], \qquad \overline{q}_j = \tfrac{1}{2}[q_j(0) + \tilde{q}_j(0)].$$

(c) Prove that, after one generation, the frequency of A_iB_j in the population, $\overline{P}_{ij} = \frac{1}{2}(P_{ij} + \tilde{P}_{ij})$, satisfies the same difference equation as the gametic frequency in a monoecious population, provided the mean recombination fraction, $\overline{c} = \frac{1}{2}(c+\tilde{c})$, is used:

$$\overline{P}'_{ij} = (1-\overline{c})\overline{P}_{ij} + \overline{c}\overline{p}_i\overline{q}_j.$$

(d) Demonstrate that

$$\overline{P}_{ij}(t) = \overline{p}_i\overline{q}_j + (1-\overline{c})^{t-1}[\overline{P}_{ij}(1) - \overline{p}_i\overline{q}_j], \qquad t \geq 1,$$

$$P_{ij}(t) = \left(\frac{1-c}{1-\overline{c}}\right)\overline{P}_{ij}(t) + \frac{1}{2}\left(\frac{c-\tilde{c}}{1-c}\right)\overline{p}_i\overline{q}_j, \qquad t \geq 2,$$

$$\tilde{P}_{ij}(t) = \left(\frac{1-\tilde{c}}{1-\overline{c}}\right)\overline{P}_{ij}(t) - \frac{1}{2}\left(\frac{c-\tilde{c}}{1-\overline{c}}\right)\overline{p}_i\overline{q}_j, \qquad t \geq 2,$$

whence $P_{ij}(t)$, $\tilde{P}_{ij}(t) \to \overline{p}_i\overline{q}_j$ as $t \to \infty$.

For two alleles at each locus, these results are due to Robbins (1918a). The generalization to multiple loci and multiple alleles was presented by Geiringer (1948) and, more concisely, by Bennett (1954).

4. SELECTION AT AN AUTOSOMAL LOCUS

We owe the basic ideas and models of selection to Fisher (1922,1930), Haldane (1924-34), and Wright (1931). After developing the multiallelic equations of selection in Section 4.1, we shall study their dynamics for two alleles in Section 4.2, and generalize to multiple alleles in Section 4.3. We shall analyze the diallelic case with a constant level of inbreeding in Section 4.4, and investigate variable environments in Section 4.5. Sections 4.6, 4.7, and 4.8 treat intra-family selection, maternal inheritance, and meiotic drive as examples of the many possible less common selection schemes. These situations require modifications of the formulation of Section 4.1. We discuss multiallelic mutation-selection balance in Section 4.9, and consider overlapping generations in Section 4.10.

4.1 *Formulation for Multiple Alleles*

We shall generalize now the fundamental model of Section 3.1 to incorporate selection. As before, we assume generations are discrete and nonoverlapping. We suppose that the sexes need not be distinguished. This will be so in a monoecious population and, for random mating, a dioecious one with the same viabilities in the two sexes and the same sex ratio in all matings. The second requirement is almost invariably satisfied. The general dioecious case is discussed in Problem 7.3.

We suppose A_iA_j zygotes have ordered frequency P_{ij}. Then the frequency of A_i in zygotes is

$$p_i = \sum_j P_{ij}. \tag{4.1}$$

Let v_{ij}, $f_{ij,kl}$, $X_{ij,kl}$ represent the viability of A_iA_j individuals and the fertility and ordered frequency of $A_iA_j \times A_kA_l$ matings. These quantities must be specified as functions of the genotypic frequencies and time, on which they may depend. Notice that the fertilities, $f_{ij,kl}$, may incorporate the effects of gametic selection (as multiplicative factors depending singly on each of the indices) and differential early survival due to variable parental care. To be included in this manner, gametic selection must act among all gametes, rather than independently among gametes in the different matings.

The frequency of A_iA_j adults is

$$\tilde{P}_{ij} = P_{ij}v_{ij}/\overline{v}, \tag{4.2}$$

where

$$\overline{v} = \sum_{ij} v_{ij}P_{ij} \tag{4.3}$$

is the mean viability of the population. Hence, the frequency of A_i in adults reads

$$\tilde{p}_i = \sum_j \tilde{P}_{ij} = p_iv_i/\overline{v}, \tag{4.4}$$

where we have introduced the viability of A_i, given by

$$p_iv_i = \sum_j P_{ij}v_{ij}. \tag{4.5}$$

It is important to note that the adults are generally not in Hardy-Weinberg proportions. We leave it for Problem 4.1 to show that if the zygotes are in Hardy-Weinberg proportions and viabilities are multiplicative (i.e., there exist c_i such that $v_{ij} = c_ic_j$), then the adults will be in Hardy-Weinberg proportions.

With random mating and no fertility differences, according to Section 3.1, the gene frequencies in zygotes in the next generation would be $\tilde{p}_i$, and the Hardy-Weinberg zygotic frequencies would be $\tilde{p}_i\tilde{p}_j$. This is the standard viability selection model. Our more general formulation will demonstrate that the form of the equation for gene frequency change is always the same, and Hardy-Weinberg proportions apply with multiplicative fertilities.

Including differential fertility in (3.6), we see that the zygotic proportions in the next generation are

$$P'_{ij} = \overline{f}^{-1} \sum_{kl} X_{ik,lj}f_{ik,lj}, \tag{4.6}$$

where

$$\overline{f} = \sum_{ijkl} X_{ik,lj}f_{ik,lj} \tag{4.7}$$

is the mean fertility. The birth rate, b_{ij}, of A_iA_j adults is defined by

$$\tilde{P}_{ij}b_{ij} = \sum_{kl} X_{ij,kl}f_{ij,kl}. \tag{4.8}$$

It is essential to distinguish the average number of offspring of an A_iA_j individual, b_{ij}, from the number of A_iA_j individuals born per mating,

$$\sum_{kl} X_{ik,lj} f_{ik,lj}.$$

The analogues of these two quantities are the same in asexual species. The fertility of an adult carrying A_i is given by

$$\tilde{p}_i b_i = \sum_j \tilde{P}_{ij} b_{ij}, \tag{4.9}$$

and, using (4.7) and (4.8), the average fecundity may be reexpressed as

$$\overline{f} = \overline{b} = \sum_{ij} b_{ij} \tilde{P}_{ij}. \tag{4.10}$$

Returning to (4.6) and employing (4.1), (4.8), (4.9), and (4.4), we find

$$p_i' = p_i (v_i b_i)/(\overline{v}\overline{b}). \tag{4.11}$$

Let us define the expected number of progeny of an $A_i A_j$ zygote,

$$w_{ij} = v_{ij} b_{ij}, \tag{4.12}$$

as the fitness of $A_i A_j$. The fitness of A_i and the mean fitness are defined by

$$p_i w_i = \sum_j P_{ij} w_{ij}, \tag{4.13}$$

$$\overline{w} = \sum_{ij} P_{ij} w_{ij}. \tag{4.14}$$

We expect

$$w_i = v_i b_i, \tag{4.15a}$$

$$\overline{w} = \overline{v}\overline{b}. \tag{4.15b}$$

To prove (4.15a), we utilize (4.13), (4.2), (4.12), (4.9), and (4.4) to derive

$$p_i w_i = \sum_j \overline{v} \tilde{P}_{ij} b_{ij} = \overline{v} \tilde{p}_i b_i = p_i v_i b_i. \tag{4.16}$$

For (4.15b), we have from (4.14), (4.13), (4.16), (4.9), and (4.10)

$$\overline{w} = \sum_i p_i w_i = \overline{v} \sum_i \tilde{p}_i b_i = \overline{v}\overline{b}.$$

Substituting (4.15) into (4.11), we deduce the basic difference equation

$$p_i' = p_i w_i / \overline{w}. \tag{4.17}$$

The change in gene frequency reads

$$\Delta p_i = p_i (w_i - \overline{w})/\overline{w}. \tag{4.18}$$

Although (4.17) has the same form as the corresponding equation, (2.6), for haploids, the fitnesses in (4.17) depend on the genotypic proportions and mating frequencies. Therefore, even if the viabilities, fertilities, and mating frequencies are specified as functions of the genotypic frequencies, (4.17) does not suffice to determine the evolution of the population. For this, we require (4.6).

To describe the dynamics entirely in terms of gene frequencies, let us posit random mating,

$$X_{ij,kl} = \tilde{P}_{ij}\tilde{P}_{kl}. \tag{4.19}$$

This reduces (4.8) to

$$b_{ij} = \sum_{kl} \tilde{P}_{kl} f_{ij,kl}. \tag{4.20}$$

We can proceed further if we confine ourselves to multiplicative fertilities (Penrose, 1949; Bodmer, 1965),

$$f_{ij,kl} = \beta_{ij}\beta_{kl}, \tag{4.21}$$

for some β_{ij}. Inserting (4.21) into (4.20) yields

$$b_{ij} = \beta_{ij}\bar{\beta}, \qquad \text{where} \qquad \bar{\beta} = \sum_{kl} \beta_{kl}\tilde{P}_{kl}, \tag{4.22}$$

and substituting this into (4.10) gives $\bar{b} = \bar{\beta}^2$, so that $\beta_{ij} = b_{ij}/\sqrt{\bar{b}}$. Therefore, (4.21) becomes

$$f_{ij,kl} = b_{ij}b_{kl}/\bar{b}. \tag{4.23}$$

Employing (4.19), (4.23), and (4.9) in (4.6) produces

$$P'_{ij} = \tilde{p}_i b_i \tilde{p}_j b_j / \bar{b}^2.$$

Substituting (4.4), (4.15), and (4.17) informs us that zygotic Hardy-Weinberg proportions,

$$P'_{ij} = p'_i p'_j, \tag{4.24}$$

are attained in a single generation.

Thus, we have derived the standard selection model,

$$P_{ij} = p_i p_j, \qquad p'_i = p_i w_i/\bar{w}, \tag{4.25a}$$

$$w_i = \sum_j w_{ij} p_j, \qquad \bar{w} = \sum_{ij} w_{ij} p_i p_j. \tag{4.25b}$$

For multiplicative fitnesses, (4.25) reduces to the haploid case, (2.6). We leave the easy proof of this assertion for Problem 4.2. Since gametic selection amounts to multiplying $f_{ij,kl}$ by factors depending on single indices, it can obviously be incorporated in (4.21). Hence, it is subsumed under (4.25) if we take $w_{ij} = \alpha_i\alpha_j\gamma_{ij}$, the factors being the gametic and zygotic fitnesses (Wright, 1969). The simplest case of (4.25) is that of constant w_{ij}, analyzed in the following two sections. The constant fitness assumption gains in plausibility if we observe that, since a genotype-independent factor cancels out of (4.25), if the $f_{ij,kl}$ are constants, β_{ij} will be constant, and consequently (4.22) permits us to use constant b_{ij} in (4.25).

If $w_{ij} = \frac{1}{2}(w_{ii} + w_{jj})$ for all i,j, we say there is no dominance. A_i is completely dominant, the other alleles being recessive to A_i, if $w_{ij} = w_{ii}$ for all j.

We can interpret (4.25) to include traits with genotypic fitnesses dependent on the environment. Suppose $x_{ij}(y)\Delta y$ is the probability that A_iA_j has a fitness between y and $y+\Delta y$. Then (4.25) applies with

$$w_{ij} = \int_0^\infty yx_{ij}(y)dy.$$

Note that if $x_{ij}(y) = x(y)$, independent of genotype, then w_{ij} is independent of genotype, and there is no selection, as expected.

If there are only two alleles, (4.18) may be rewritten in the form

$$\Delta p = pq(w_1-w_2)/\overline{w}, \qquad (4.26)$$

where, as usual, $p = p_1$ and $q = p_2$. It is often convenient to set $w_{11} = 1+s$, $w_{12} = 1+hs$, $w_{22} = 1-s$. Then s represents the intensity of selection; h specifies the degree of dominance: $h = 0$ if there is no dominance and $h = 1$ (-1) if A_1 is dominant (recessive). We designate the situations $w_{12} > w_{11},w_{22}$ $(h > 1)$ and $w_{12} < w_{11},w_{22}$ $(h < -1)$ by the terms *overdominance* and *underdominance*.

4.2 *Dynamics with Two Alleles*

The difference equation (4.26) with constant fitnesses is known to be solvable only in two cases. If fitnesses are multiplicative, as mentioned above, it reduces to the haploid case, and this was solved for any number of alleles in Section 2.1. We can also calculate $p(t)$ if one of the homozygotes has zero fitness (Problem 4.3). This can happen either because the allele involved, say a $(= A_2)$, is *lethal*, which means aa is inviable $(v_{22} = 0)$, or because aa is *sterile* $(b_{22} = 0)$. We shall determine the fate of the population for all possible fitnesses (Fisher, 1922, 1930). This global analysis is a special case of that in Section 4.4; it is given separately in order to present the fundamental ideas below in their simplest form. We shall also calculate the rates of convergence to equilibrium for the various fitness patterns.

We rewrite (4.26) in the form

$$\Delta p = pqg(p)/\bar{w}, \tag{4.27}$$

where

$$g(p) = w_{12}-w_{22} + (w_{11}-2w_{12}+w_{22})p. \tag{4.28}$$

We exclude the trivial case of no selection by assuming the w_{ij} are not all equal. There are four distinct cases, depending on the signs of

$$g(0) = w_{12}-w_{22} \quad \text{and} \quad g(1) = w_{11}-w_{12}.$$

1. $w_{22} \leq w_{12} \leq w_{11}$

We have $g(0) \geq 0$ and $g(1) \geq 0$, so the linear function $g(p) > 0$ for $0 < p < 1$. Since $\operatorname{sgn} \Delta p = \operatorname{sgn} g(p)$, therefore $p(t) \to 1$ as $t \to \infty$. This is not surprising since A is favored over a for all gene frequencies.

2. $w_{11} \leq w_{12} \leq w_{22}$

This is just Case 1 with the sign of $g(p)$ reversed. Therefore, $p(t) \to 0$ as $t \to \infty$. The deleterious allele is again eliminated.

3. $w_{12} < w_{11}, w_{22}$

In the underdominant case, we have an equilibrium

$$\hat{p} = \frac{w_{22}-w_{12}}{w_{11}-2w_{12}+w_{22}} \tag{4.29}$$

in (0,1). From (4.27), (4.28), and (4.29) we obtain

$$\Delta p = (w_{11}-2w_{12}+w_{22})pq(p-\hat{p}) / \bar{w}, \tag{4.30}$$

whence $\operatorname{sgn} \Delta p = \operatorname{sgn} (p-\hat{p})$. Consequently, the equilibrium $\hat{p}$ is unstable: $p(t) \to 0$ if $p(0) < \hat{p}$ and $p(t) \to 1$ if $p(0) > \hat{p}$. Thus, the fate of the population depends on its initial condition. This is frequently the situation with an unstable polymorphism.

4. $w_{12} > w_{11}, w_{22}$

The equilibrium $\hat{p}$ is still in (0,1) in the overdominant case, but now (4.30) informs us that $\operatorname{sgn} \Delta p = \operatorname{sgn} (\hat{p}-p)$. Due to the *a priori* possibility of diverging oscillation about $\hat{p}$, this result does not prove that $\hat{p}$ is stable. Let us show that $p(t) \to \hat{p}$ without oscillation, provided $0 < p(0) < 1$. Since our method will be used repeatedly, we shall describe it quite generally.

Suppose the continuous mapping $p' = f(p)$ of $[a,b]$ into itself has a fixed point $\hat{p} = f(\hat{p})$ in (a,b). We define $x = p-\hat{p}$, and assume the mapping can be rewritten in the form $x' = \lambda(p)x$, with $0 \leq \lambda(p) < 1$ for $a < p < b$. Then $\hat{p}$ is the only fixed point in (a,b), and $p(t) \to \hat{p}$ as $t \to \infty$ without oscillation.

The proof is easy. If $\bar{p} \neq \hat{p}$ is a fixed point in (a,b), then $\bar{x} = \bar{p}-\hat{p} \neq 0$, so $\lambda(\bar{p}) = 1$, contradicting $\lambda(p) < 1$. To prove convergence, note that $\{|x(t)|\}$ is a monotone decreasing sequence of positive numbers. Hence, it tends to a limit. By our uniqueness argument, this limit must be zero. Since $\lambda(p) > 0$, the convergence is nonoscillatory.

We give a more general theorem of this type in Problem 4.5.

To apply the above result to our problem, it is convenient to choose $w_{11} = 1-r$, $w_{12} = 1$, $w_{22} = 1-s$, with $0 < r,s \leq 1$. In terms of the selection coefficients r and s, (4.29) and (4.30) become

$$\hat{p} = \frac{s}{r+s}, \tag{4.31}$$

$$\Delta p = -(r+s)pqx/\bar{w}, \tag{4.32}$$

where $x = p-\hat{p}$ and

$$\bar{w} = 1-rp^2-sq^2. \tag{4.33}$$

Now,

$$x' = x+\Delta x = x+\Delta p = \lambda(p)x, \tag{4.34}$$

where (4.32) yields

$$\lambda(p) = 1-\frac{(r+s)pq}{\bar{w}}. \tag{4.35}$$

Obviously, for $0 < p < 1$, $\lambda(p) < 1$. Substituting (4.33) into (4.35) gives

$$\lambda(p) = \frac{1-rp-sq}{1-rp^2-sq^2}. \tag{4.36}$$

Since $0 < r,s \leq 1$, therefore

$$rp^2+sq^2 < rp+sq \leq 1, \qquad 0 < p < 1,$$

and (4.36) shows that $\lambda(p) \geq 0$.

We conclude that the overdominant polymorphism is globally

stable. This is one of the basic mechanisms for maintaining genetic variability. Observe that it has no analogue in a haploid population.

The example of sickle-cell anemia in Negroes is discussed in detail by Cavalli-Sforza and Bodmer (1971). Briefly, AA individuals are normal, Aa is slightly anemic, having both normal and abnormal hemoglobin, but is probably more resistant to malaria, and aa, having only abnormal hemoglobin, has severe anemia. Typical selection coefficients in African populations are $r \approx 0.15$ and $s \approx 1$, whence $\hat{q} \approx 0.13$.

We shall now calculate the rates of approach to equilibrium. Clearly, it will suffice to investigate $p(t) \to \hat{p}$ and $p(t) \to 0$.

(i) $p(t) \to \hat{p}$

This is Case 4 above. As $p \to \hat{p}$, (4.34) simplifies to $x' \approx \lambda(\hat{p})x$, whence we obtain as $t \to \infty$ convergence at the geometric rate

$$x(t) \approx (\text{const})[\lambda(\hat{p})]^t = (\text{const})\left(\frac{r+s-2rs}{r+s-rs}\right)^t. \tag{4.37}$$

For weak selection, this rate is approximately

$$x(t) \approx (\text{const})\exp\left(-\frac{rst}{r+s}\right), \qquad r,s \ll 1,$$

in which case the characteristic convergence time is $(r+s)/(rs)$.

See Problem 4.6 for the calculation of corrections to geometric rates like (4.37).

(ii) $p(t) \to 0$

We can have either Case 2 or Case 3 above. It is necessary to separate the case of recessive A.

(a) $w_{12} < w_{22}$

As $p \to 0$, the exact equation

$$p' = pw_1/\bar{w} \tag{4.38}$$

reduces to $p' \approx pw_{12}/w_{22}$, which yields the geometric rate

$$p(t) \approx (\text{const})(w_{12}/w_{22})^t, \qquad t \to \infty.$$

If

$$\frac{w_{12}}{w_{22}} = 1-k,$$

for weak selection ($k \ll 1$) the rate is roughly $(\text{const})\,e^{-kt}$.

(b) $w_{12} = w_{22} > w_{11}$

To examine the elimination of a recessive, we must approximate (4.38) to second order in p. The failure of the linear approximation leading to a geometric rate is not surprising: selection is only against AA, and the frequency of these individuals, p^2, is extremely small for $p \ll 1$. Therefore, we expect a very slow rate of ultimate elimination of the recessive allele A. With $w_{11}=1-\sigma$, $w_{12}=w_{22}=1$, (4.38) reads

$$p' = \frac{p(1-\sigma p)}{1-\sigma p^2}. \tag{4.39}$$

It is helpful to rewrite (4.39) in terms of a new variable, y, with the property $\Delta y \to$ const as $p \to 0$. We require $y = 1/p$, for which

$$y' = y\frac{1-(\sigma/y^2)}{1-(\sigma/y)}.$$

Therefore,

$$\Delta y = \sigma - \sigma(1-\sigma)y^{-1} + O(y^{-2}) \tag{4.40}$$

as $y \to \infty$. Omitting the terms of $O(y^{-2})$ in (4.40) will be a good approximation for sufficiently long times: $t \geq T$, for some fixed $T \gg 1$. But

$$y(t) = y(T) + \sum_{\tau=T}^{t-1} [y(\tau+1)-y(\tau)]$$

$$= \sigma t + \text{const} + \sum_{\tau=T}^{t-1} \left[-\frac{\sigma(1-\sigma)}{y(\tau)} + O(y^{-2})\right]$$

from (4.40). The dominant contribution to the sum must come from the larger y^{-1} term, and $y \approx \sigma t$ for $t \gg 1$. Since

$$\sum_{\tau=T}^{t-1} \tau^{-1} = \ln t + O(1)$$

as $t \to \infty$, we obtain

$$y(t) = \sigma t - (1-\sigma)\ln t + O(1)$$

as $t \to \infty$. Therefore

$$p(t) = \frac{1}{\sigma t} + \frac{(1-\sigma)\ln t}{\sigma^2 t^2} + O(t^{-2}). \tag{4.41}$$

The algebraic decay (4.41) is quite slow: it takes about $100/\sigma$ generations to reduce $p(t)$ to 0.01; the corresponding time for exponential convergence would be about $4.6/\sigma$. It is at first sight puzzling to observe that, in contrast to the leading term for geometric convergence, the first two terms in (4.41) have no arbitrary constant corresponding to $p(T)$. This happens because to this order t may be replaced by $t + \text{const}$.

4.3 *Dynamics with Multiple Alleles*

We shall now analyze (4.25) with constant w_{ij} and k alleles. Let us investigate first the existence of polymorphic equilibria (Mandel, 1959; Hughes and Seneta, 1975). If $\hat{p}_i = 0$ for some set of subscripts, we just have a system of lower dimension, so we may assume all the alleles are present: $\hat{p}_i > 0$ for all i. Then (4.25) gives the equilibrium condition $\hat{w}_i = \hat{\bar{w}}$. Writing W for the $k \times k$ matrix of the w_{ij} and $\underset{\sim}{u}$ for the $k \times 1$ vector of ones, we obtain $W\underset{\sim}{\hat{p}} = \hat{\bar{w}}\underset{\sim}{u}$. This relation becomes more illuminating with the introduction of $\operatorname{adj} W$, the adjoint matrix of W; $\det W$, the determinant of W; and I, the $k \times k$ identity matrix. These entities satisfy

$$(\operatorname{adj} W)W = (\det W)I. \tag{4.42}$$

Multiplying our vector equilibrium equation by $\operatorname{adj} W$ and using (4.42) yields

$$(\det W)\,\underset{\sim}{\hat{p}} = \hat{\bar{w}}\,(\operatorname{adj} W)\,\underset{\sim}{u}.$$

We denote the ith component of the vector $(\operatorname{adj} W)\underset{\sim}{u}$ by W_i, i.e., $W_i = [(\operatorname{adj} W)\underset{\sim}{u}]_i$. Hence,

$$(\det W)\hat{p}_i = \hat{\bar{w}}W_i, \tag{4.43}$$

If $\det W \neq 0$, $W_i \neq 0$ for any i, and all the W_i have the same sign, then there exists a unique internal equilibrium

$$\hat{p}_i = \frac{\hat{\bar{w}}W_i}{\det W} = \frac{W_i}{\sum_j W_j}. \tag{4.44}$$

The second equation follows by substituting for $\det W$ from

$$\det W = \hat{\bar{w}} \sum_i W_i,$$

which just normalizes the allelic frequencies in (4.43). We can also see from (4.44) that if $\det W \neq 0$ and an internal equilibrium exists, then it is unique, $W_i \neq 0$ for any i, and all the W_i have the same sign.

The degenerate case, $\det W = 0$, is treated by Hughes and Seneta (1975).

If the fitness matrix is completely nondegenerate, the maximum possible number of equilibria is $2^k - 1$. To see this, choose any subset

S of the integers $1,2,\ldots,k$. The principal minors of W are the matrices W_S formed from the rows and columns corresponding to the elements of S. We suppose each of these principal minors has a nonzero determinant. Then the number of equilibria with the alleles corresponding to the elements of S present, and only those alleles present, is 0 or 1. Since each of the k alleles must be present or absent, but at least one must be present, the number of possible equilibria is 2^k-1.

If at least one of the principal minors W_S has determinant zero, then lines, surfaces, etc. of equilibria are possible. The most degenerate case is no selection, w_{ij} independent of i and j, for which every point in the *simplex* $p_i \geq 0$, $\sum_i p_i = 1$ is an equilibrium.

We shall now show that the stationary points of the mean fitness are equilibria of the population and *vice versa* (Mandel, 1959). We put

$$f(\underset{\sim}{p}) = \overline{w}(\underset{\sim}{p}) - \lambda \sum_i p_i,$$

where λ is a Lagrange multiplier. Therefore,

$$\frac{\partial f}{\partial p_i} = 2w_i - \lambda = 0$$

at the stationary points. But

$$0 = \sum_i p_i \frac{\partial f}{\partial p_i} = 2\overline{w} - \lambda.$$

So the stationary points are precisely the points where $w_i = \overline{w}$, i.e., the equilibria.

Let us demonstrate now that the mean fitness is nondecreasing, and the change in mean fitness is zero only at equilibrium. This extremely important and useful result has been proved by Scheuer and Mandel (1959); Mulholland and Smith (1959); Atkinson, Watterson, and Moran (1960); and, most elegantly, Kingman (1961), whom we follow.

We shall require Jensen's inequality: For a convex function g, defined on an open interval I, and a random variable x, with a distribution $q(x)$ concentrated on I,

$$E[g(x)] \geq g[E(x)], \tag{4.45}$$

with equality if and only if $g(x)$ is linear or $q(x)$ is concentrated at a point. For a concave function, the inequality is reversed.

The proof is immediate. If $g(x)$ is convex, at every point $P[\xi,g(\xi)]$ on its graph, there exists a line lying entirely below or on $g(x)$. Therefore,

$$g(x) \geq g(\xi) + m(x-\xi),$$

for all x in I, where m is the slope of the line. Choosing $\xi = E(x)$ and taking expectations, we obtain (4.45).

If x takes values x_i with probabilities p_i, (4.45) becomes

$$\sum_i p_i g(x_i) \geq g\left(\sum_i p_i x_i\right). \tag{4.46}$$

For $g(x) = x^\mu$, $\mu \geq 1$, (4.46) reads

$$\sum_i p_i x_i^\mu \geq \left(\sum_i p_i x_i\right)^\mu, \tag{4.47}$$

the form we shall use in the ensuing proof.

We write a series of inequalities for the mean fitness in the next generation, and explain the manipulations below.

$$\overline{w}' = \sum_{ij} p_i' p_i' w_{ij}$$

$$= \overline{w}^{-2} \sum_{ij} p_i p_j w_i w_j w_{ij} \tag{4.48}$$

$$= \overline{w}^{-2} \sum_{ijk} p_i p_j p_k w_{ij} w_{ik} (1/2)(w_j + w_j) \tag{4.49}$$

$$= \overline{w}^{-2} \sum_{ijk} p_i p_j p_k w_{ij} w_{ik} (1/2)(w_j + w_k)$$

$$\geq \overline{w}^{-2} \sum_{ijk} p_i p_j p_k w_{ij} w_{ik} (w_j w_k)^{1/2} \tag{4.50}$$

$$= \overline{w}^{-2} \sum_i p_i \left[\sum_j p_j w_{ij} (w_j)^{1/2} \right]^2$$

$$\geq \overline{w}^{-2} \left[\sum_i p_i \sum_j p_j w_{ij} (w_j)^{1/2} \right]^2 \tag{4.51}$$

$$= \overline{w}^{-2} \left(\sum_j p_j w_j^{3/2} \right)^2 \tag{4.52}$$

$$\geq \overline{w}^{-2} \left[\left(\sum_j p_j w_j \right)^{3/2} \right]^2 \tag{4.53}$$

$$= \overline{w}.$$

The numbered equations in the above series come, respectively, from (4.25a), (4.25b), the elementary fact that $a+b \geq 2(ab)^{1/2}$ for $a,b \geq 0$, (4.47) with $\mu = 2$, (4.25b), and (4.47) with $\mu = 3/2$.

Notice how much easier it was to prove $\Delta\overline{w} \geq 0$ for haploids in Section 2.1. We can gain considerable evolutionary insight by following the approach in that section to evaluate $\Delta\overline{w}$. For the purpose of this computation, we shall permit the fitnesses to vary. We have

$$\Delta\overline{w} = \sum_{ij} (p'_i p'_j w'_{ij} - p_i p_j w_{ij})$$

$$= \overline{\Delta w} + \sum_{ij} [2p_j \Delta p_i + \Delta p_i \Delta p_j] w_{ij}, \qquad (4.54)$$

where

$$\overline{\Delta w} = \sum_{ij} p'_i p'_j \Delta w_{ij} \qquad (4.55)$$

is the mean of the fitness changes over the next generation. Substituting (4.18) into (4.54) leads to

$$\Delta\overline{w} = \overline{w}^{-1} V_g + \overline{\Delta w} + \overline{w}^{-2} \sum_{ij} p_i p_j (w_i - \overline{w})(w_j - \overline{w})(w_{ij} - \overline{w}), \qquad (4.56)$$

where

$$V_g = 2 \sum_i p_i (w_i - \overline{w})^2 \qquad (4.57)$$

is the *genic* (or *additive genetic*) variance in fitness. It will be shown in Section 4.10 (4.179 to 4.189) that V_g is, in fact, the additive component of the *genotypic* (or *total genetic*) variance. The factor 2 in (4.57) is due to diploidy. Note, too, that the last term in (4.56) was absent in the haploid result, (2.16).

It is very instructive to examine (4.56) for weak selection. Let us define more formally the symbol O we have already introduced. The notation

$$\phi(x) = O[\psi(x)] \quad \text{as } x \to a$$

means

$$\overline{\lim_{x \to a}} \, |\phi(x)/\psi(x)| < \infty,$$

i.e., $|\phi(x)/\psi(x)|$ is bounded for $x \neq a$, but x sufficiently close to a. If

$$\lim_{x \to a} |\phi(x)/\psi(x)| = 0,$$

we shall write

$$\phi(x) = o[\psi(x)] \quad \text{as } x \to a.$$

Of course, this means $|\phi(x)/\psi(x)|$ can be made arbitrarily small for $x \neq a$, but x sufficiently close to a.

Since the scale of w_{ij} is irrelevant, we may take

$$w_{ij} = 1 + O(s) \qquad \text{as } s \to 0, \tag{4.58}$$

where s is the intensity of selection. For instance, if the genotypic fitnesses are constant, we can choose s to be the largest absolute selective difference:

$$s = \max_{ijkl} |w_{ij} - w_{kl}|.$$

Now, (4.57) shows that $V_g = O(s^2)$, and the last term in (4.56) is of $O(s^3)$. Since (4.58) implies $\overline{w} = 1 + O(s)$, (4.56) has the form

$$\Delta\overline{w} = V_g + \overline{\Delta w} + O(s^3). \tag{4.59}$$

If the w_{ij} are constant or change so slowly that $\overline{\Delta w} = o(s^2)$, then we have the approximate form of the Fundamental Theorem of Natural Selection, $\Delta\overline{w} \approx V_g$. Thus, as the mean fitness increases (proved exactly above), it does so at a rate approximately equal to the additive component of the genetic variance. From (4.25a) and (4.57) we infer that $V_g = 0$ if and only if the population is at equilibrium.

The stability properties of the basic selection model were studied by Kimura (1956) and Penrose, S. Maynard Smith, and Sprott (1956) in continuous time. The discrete model has been investigated by Mandel (1959) and Kingman (1961a). Mandel (1970) and Hughes and Seneta (1975) examine the relations among the different stability criteria. The following results are either proved in or can be deduced from Kingman's (1961a) paper.

1. The stable equilibria are the local maxima of $\overline{w}$. (These correspond to the peaks, ridges, etc. of the $\overline{w}$ surface, which the system climbs because $\Delta\overline{w} \geq 0$.) The asymptotically stable equilibria are the strict (or isolated) local maxima of $\overline{w}$.

   Let S be a subset of the integers $1,2,\ldots,k$. Suppose there exists an equilibrium, $\hat{\underset{\sim}{p}}(S)$, with $\hat{p}_i \neq 0$ if and only if i is in S. Then (4.44) tells us that $\hat{\underset{\sim}{p}}(S)$ is isolated if and only if the principal minor W_S, corresponding to S, has nonzero determinant.

2. If the completely polymorphic equilibrium (i.e., with all alleles present) exists, it is stable if and only if, counting multiplicity, the fitness matrix W has exactly one positive eigenvalue. If $\det W \neq 0$, by Result 1, stability is necessarily asymptotic.

3. Let D_i, $i = 1,2,\ldots,k$, be the determinant of the $i \times i$ submatrix in the upper left-hand corner of W. Suppose, without loss of biological generality, $w_{11} = D_1 > 0$, and assume $D_i \neq 0$ for any i. Then

the stability of the completely polymorphic equilibrium is necessarily asymptotic, and a necessary and sufficient condition for this stability is

$$(-1)^i D_i < 0 \qquad \text{for all } i.$$

Of course, this criterion is a special case of Result 2; we state it because it is much easier to apply.

4. The fitness of A_i and the mean fitness at $\hat{\underset{\sim}{p}}(S)$ are

$$\hat{w}_i(S) = \sum_j w_{ij}\hat{p}_j(S),$$

$$\hat{\bar{w}}(S) = \sum_{ij} w_{ij}\hat{p}_i(S)\hat{p}_j(S).$$

The equilibrium $\hat{\underset{\sim}{p}}(S)$ is stable if and only if

$$\hat{w}_i(S) \le \hat{\bar{w}}(S) \qquad \text{for all } i \text{ not in } S,$$

and W_S has exactly one positive eigenvalue. Observe that these criteria require that alleles not present at equilibrium die out if introduced at low frequencies, and that the alleles present be in stable equilibrium (see Result 2).

5. If W has m positive eigenvalues, at least m-1 alleles must be absent at a stable equilibrium.

6. Let S and T be distinct subsets of $\{1,2,\ldots,k\}$ such that T is contained in S. Suppose $\theta\hat{\underset{\sim}{p}}(S) + (1-\theta)\hat{\underset{\sim}{p}}(T)$ is *not* an equilibrium for any θ in $(0,1)$. Then the equilibria $\hat{\underset{\sim}{p}}(S)$ and $\hat{\underset{\sim}{p}}(T)$ cannot both be stable. Hence, if $\hat{\underset{\sim}{p}}(S)$ is stable, and $\hat{\underset{\sim}{p}}(S)$ and $\hat{\underset{\sim}{p}}(T)$ *cannot* be connected by a line of equilibria, then $\hat{\underset{\sim}{p}}(T)$ is unstable. This result often eliminates much calculation in testing the stability of equilibria.

In particular, if the completely polymorphic equilibrium is stable, then no other equilibrium is stable, and therefore (provided all alleles are initially present) the population converges globally to the completely polymorphic equilibrium. Note that if this equilibrium is degenerate, it will include boundary points, lines, etc., with at least one allele absent, and these boundary equilibria are stable. The population converges to these boundary equilibria only if at least one allele is initially absent.

7. Assume the population is in the stable completely polymorphic equilibrium $\hat{\underset{\sim}{p}}$. The mean fitness at this equilibrium is

$$\hat{\bar{w}} = \sum_{i,j=1}^{k} w_{ij}\hat{p}_i\hat{p}_j,$$

and a new allele A_{k+1} has equilibrium fitness

$$\hat{w}_{k+1} = \sum_{i=1}^{k} w_{k+1,i}\hat{p}_i.$$

It follows from Results 4 and 6 that a necessary and sufficient condition for A_{k+1} to persist in the population (i.e., for a new stable equilibrium with A_{k+1} present to be eventually established) is $\hat{w}_{k+1} > \hat{\bar{w}}$.

4.4 *Two Alleles with Inbreeding*

We return to the general model, (4.13), (4.14), (4.17), and consider two alleles with inbreeding. We must specify the genotypic proportions, P_{ij}, in terms of the allelic frequencies, p_i. The required relations read

$$P_{11} = p^2 + Fpq, \quad P_{12} = (1-F)pq, \quad P_{22} = q^2 + Fpq. \tag{4.60}$$

For the purpose of this section, the inbreeding coefficient F, $0 \leq F \leq 1$, is just a parameter that determines the excess of homozygotes and deficiency of heterozygotes relative to Hardy-Weinberg proportions. Observe that (4.60) satisfies (4.1). Assuming that F is constant obviates explicit consideration of the mating system; an example of the latter approach is treated in Section 5.1.

The model with constant inbreeding coefficient is due to Wright (1942). Assuming a polymorphism exists, Lewontin (1958) has given conditions for its local stability. We shall determine globally the fate of the population for all values of the fitnesses and the inbreeding coefficient. The method is that of Section 4.2.

Substituting (4.60) into (4.13), we calculate w_1 and w_2, and hence $\bar{w} = pw_1 + qw_2$. Then (4.26) still has the form (4.27), but now

$$g(p) = \beta - F\alpha - (1-F)\gamma p, \tag{4.61}$$

$$\bar{w} = w_{11}p^2 + (2w_{12} - F\gamma)pq + w_{22}q^2, \tag{4.62}$$

where

$$\alpha = w_{12} - w_{11}, \qquad \beta = w_{12} - w_{22}, \tag{4.63}$$

$$\gamma = \alpha+\beta = 2w_{12}-w_{11}-w_{22}. \tag{4.64}$$

We remark that $\alpha = 0$, $\beta = 0$, and $\gamma = 0$ correspond to A dominant, A recessive, and no dominance. Since $\operatorname{sgn} \Delta p = \operatorname{sgn} g(p)$, and $g(p)$ is again linear, as in Section 4.2, there are four cases depending on the signs of

$$g(0) = \beta-F\alpha \quad \text{and} \quad g(1) = -\alpha+F\beta. \tag{4.65}$$

We exclude two trivial situations: no selection ($w_{11} = w_{12} = w_{22}$) and complete inbreeding ($F = 1$). The latter is easily seen from (4.27) and (4.61) to be equivalent to the diallelic haploid model (see Section 2.1) with $w_1 \to w_{11}$ and $w_2 \to w_{22}$. The fitness of Aa individuals becomes irrelevant because there are no heterozygotes in the population.

1. $\beta \geq F\alpha$ and $F\beta \geq \alpha$

Since $\Delta p > 0$ for $0 < p < 1$, therefore $p(t) \to 1$ as $t \to \infty$. If $w_{11} \geq w_{12} \geq w_{22}$, then $\alpha \leq 0$ and $\beta \geq 0$, so this case necessarily applies for all F, as expected.

2. $\beta \leq F\alpha$ and $F\beta \leq \alpha$

Now $\Delta p < 0$ for $0 < p < 1$, so $p(t) \to 0$. If $w_{11} \leq w_{12} \leq w_{22}$, then $\alpha \geq 0$ and $\beta \leq 0$, whence the requirements for $p(t) \to 0$ hold for all F.

3. $\beta < F\alpha$ and $F\beta > \alpha$

There is a polymorphism

$$\hat{p} = \frac{\beta-F\alpha}{\gamma(1-F)} \tag{4.66}$$

in $(0,1)$; and (4.27), (4.61), and (4.66) give

$$\Delta p = \gamma(1-F)pq(\hat{p}-p)/\bar{w}. \tag{4.67}$$

Since

$$0 > g(0)-g(1) = \gamma(1-F), \tag{4.68}$$

therefore $\operatorname{sgn} \Delta p = \operatorname{sgn}(p-\hat{p})$. Hence, $p(t) \to 1$ if $p(0) > \hat{p}$ and $p(t) \to 0$ if $p(0) < \hat{p}$.

We solve (4.65) for α and β to obtain

$$\alpha = \frac{Fg(0)-g(1)}{1-F^2}, \qquad \beta = \frac{g(0)-Fg(1)}{1-F^2}. \tag{4.69}$$

Thus, the conditions for Case 3, $g(0) < 0$ and $g(1) > 0$, imply $\alpha,\beta < 0$, i.e., $w_{12} < w_{11},w_{22}$. Consequently, this unstable equilibrium requires underdominance for all F. In contrast to panmixia, however, underdominance need not lead to an unstable polymorphism.

4. $\beta > F\alpha$ and $F\beta < \alpha$

The nontrivial equilibrium $\hat{p}$ still exists, but now $\operatorname{sgn} \Delta p = \operatorname{sgn}(\hat{p} - p)$. We expect $p(t) \to \hat{p}$ globally. Before proving this, we note from the previous paragraph that our conditions now require overdominance for all F, but overdominance is not sufficient for a stable polymorphism.

We put $x = p - \hat{p}$ and obtain (4.34) with

$$\lambda(p) = 1 - \frac{\gamma(1-F)pq}{\bar{w}}. \tag{4.70}$$

Since the inequality in (4.68) is reversed here, it follows at once that $\lambda(p) < 1$, $0 < p < 1$.

To prove that $\lambda(p) \geq 0$, we take $w_{12} = 1$. Then $0 < \alpha, \beta \leq 1$. Substituting (4.63) into (4.62) leads to

$$\bar{w} = 1 - \alpha p^2 - F\gamma pq - \beta q^2,$$

and inserting this into (4.70) yields

$$\bar{w}\lambda(p) = 1 - \alpha p - \beta q \geq 0.$$

This proves global nonoscillating convergence.

4.5 *Variable Environments*

We shall investigate in this section the effect of deterministic time-dependence of the fitnesses. We assume that the $w_{ij}(t)$ are known functions of time, and restrict ourselves to two alleles. With arbitrary $w_{ij}(t)$, not only is a global analysis impossible, but we cannot even locate the possible equilibria. We can, however, still find sufficient conditions for the maintenance of genetic diversity. If, no matter what the initial conditions are, the population cannot become monomorphic, we shall say there is a *protected polymorphism* (Prout, 1968). In general, a protected polymorphism is neither necessary nor sufficient for the existence of a stable polymorphic equilibrium. For instance, on the one hand, there may be a (locally) stable polymorphic equilibrium, but selection may remove a rare allele from the population. On the other hand, polymorphism may be maintained, e.g., by a stable limit cycle to which there is global convergence.

To obtain sufficient conditions for a protected polymorphism, clearly, we must examine (4.25) near $p = 0$ and $q = 0$. As $p \to 0$, (4.25a) reduces to

$$p(t+1) = p(t)w_{12}(t)/w_{22}(t), \tag{4.71}$$

which is immediately iterated:

$$p(t) = p(0)\prod_{\tau=0}^{t-1} \frac{w_{12}(\tau)}{w_{22}(\tau)}. \tag{4.72}$$

We want p to increase when A is rare. From (4.72) we infer that the proper requirement is $w_{12}^*(t) > w_{22}^*(t)$ for sufficiently large t, where

$$w_{ij}^*(t) = \left[\prod_{\tau=0}^{t-1} w_{ij}(\tau)\right]^{1/t} \tag{4.73}$$

is the geometric mean. Adding the corresponding condition for q to increase when a is rare, obtained at once by interchanging the subscripts 1 and 2, we obtain geometric mean overdominance,

$$w_{12}^*(t) > w_{11}^*(t), w_{22}^*(t), \tag{4.74}$$

for sufficiently large t.

For (4.74) to suffice for a protected polymorphism, we must make certain $p(t)$ cannot reach 0 or 1 in a single step from the interior. Then we cannot have, e.g., sequences $\{p(2t)\}$ and $\{p(2t+1)\}$ with $p(2t)$ not close to 0 or 1, but $p(2t+1) \to 0$ as $t \to \infty$, because $\{p(2t)\}$ would have a point of accumulation in the interior, from which the population would move to 0 in a single generation. If we stipulate that the heterozygote fitness, $w_{12}(t)$, cannot vanish simultaneously with either homozygote fitness, (4.25) informs us that $0 < p(t) < 1$ implies $0 < p(t+1) < 1$.

Writing

$$\overline{w}_{ij}(t) = \frac{1}{t}\sum_{\tau=0}^{t-1} w_{ij}(\tau) \tag{4.75}$$

for the arithmetic mean, we have (Problem 4.10)

$$w_{ij}^*(t) \le \overline{w}_{ij}(t). \tag{4.76}$$

It is easy to see that (4.74) is more or less stringent than arithmetic mean overdominance according as the heterozygote or homozygote fitnesses are more variable. With constant homozygote fitnesses, (4.74) becomes

$$w_{12}^*(t) > w_{11}, w_{22},$$

which, in view of (4.76), is stronger than

$$\overline{w}_{12}(t) > w_{11}, w_{22}.$$

If $w_{12}(t)$ is constant, however, (4.74) reads

$$w_{11}^*(t), w_{22}^*(t) < w_{12},$$

and this can be satisfied without arithmetic mean overdominance.

For the sickle-cell anemia example of Section 4.2, the abnormal homozygotes are quite unfit, $w_{22}(t) << w_{12}(t)$. If heterozygotes were generally slightly less fit than normal homozygotes, but occasional malaria epidemics eliminated a large proportion of normal homozygotes, we would have $\overline{w}_{11} > \overline{w}_{12}$ and $w_{11}^* < w_{12}^*$. Therefore, (4.74) would be satisfied, and the allele a for sickling would be maintained in the population.

Equation (4.74) is due to Haldane and Jayakar (1963). If the environment is periodic, we may calculate w_{ij}^* over one period; with $w_{ij}(t+T) = w_{ij}(t)$, the period would be T generations. Such a model may apply to seasonal variations in many organisms. Cyclic environments are analyzed in detail by Karlin and Lieberman (1974), Hoekstra (1975), and Nagylaki (1975).

If a is dominant, $w_{12}(t) = w_{22}(t)$, so (4.74) is violated. We still insist on $w_{12}^*(t) > w_{11}^*(t)$, but now must expand (4.25a) to second order to obtain a condition near $p = 0$. With $y(t) = 1/p(t)$, $w(t) = w_{11}(t)/w_{12}(t)$, and $\sigma(t) = 1-w(t)$, (4.40) reads

$$y(t+1) - y(t) = \sigma(t) + O[1/y(t)],$$

whence

$$y(t) - y(0) \sim \sum_{\tau=0}^{t-1} \sigma(\tau).$$

Thus, A will not be lost if

$$\sum_{\tau=0}^{t-1} \sigma(\tau) = t - \sum_{\tau=0}^{t-1} w(\tau) = t[1 - \overline{w}(t)] < 0,$$

i.e., $\overline{w}(t) > 1$, for sufficiently large t. In order not to lose an allele in a single generation, we require $w_{12}(t) \neq 0$ for any t. Then the sufficient conditions for protection become

$$w^*(t) < 1 \quad \text{and} \quad \overline{w}(t) > 1 \qquad (4.77)$$

for sufficiently large t (Haldane and Jayakar, 1963).

Unlike the geometric mean requirement, the arithmetic mean condition on the relative recessive fitness cannot be expressed as a comparison between the means of absolute fitnesses. We expect (4.77) to be satisfied if the recessive is usually slightly favored, but is severely deleterious on rare occasions. For cyclic environments, the reader may consult Dempster (1955), Karlin and Lieberman (1974), and Hoekstra (1975).

4.6 *Intra-Family Selection*

Let us suppose that the number of adults contributed by a family to the next generation is independent of the parental genotypes. Within families, however, A_iA_j individuals have a relative viability x_{ij}. Therefore, the contribution of A_iA_j adults of an $A_iA_k \times A_lA_j$ mating is proportional to $x_{ij}c_{ik,lj}$, where

$$c_{ik,lj} = 4(x_{ij} + x_{il} + x_{kl} + x_{kj})^{-1}. \tag{4.78}$$

For random mating, our basic equation reads

$$P'_{ij} = \sum_{kl} P_{ik}P_{lj}x_{ij}c_{ik,lj}, \tag{4.79}$$

where P_{ij} is the frequency of A_iA_j adults.

To verify that the normalization in (4.79) is correct, note that

$$c_{ik,lj} = c_{ki,lj} = c_{lj,ik}, \tag{4.80}$$

giving us

$$\begin{aligned}\sum_{ij} P'_{ij} &= (1/4)\sum_{ijkl} P_{ik}P_{lj}(x_{ij} + x_{il} + x_{kl} + x_{kj})c_{ik,lj} \\ &= \sum_{ijkl} P_{ik}P_{lj} \\ &= 1.\end{aligned}$$

To study weak selection, let us put $x_{ij} = 1 + s_{ij}$, the intra-family selection coefficients, s_{ij}, being small. From (4.79) we obtain at once

$$P'_{ij} = \sum_{kl} P_{ik}P_{lj} + O(s) = p_ip_j + O(s), \tag{4.81}$$

i.e., the deviation of the adults from Hardy-Weinberg proportions is reduced to the order of the selection intensity in one generation of panmixia. The zygotes, of course, are exactly in Hardy-Weinberg ratios, but, since the selection is mating-dependent, this fact is not useful in the formulation or analysis. The gene frequency change may be computed by using (4.78), (4.79), and (4.80):

$$\Delta p_i = \sum_{jkl} P_{ik}P_{lj}(x_{ij}c_{ik,lj} - 1)$$

$$= \frac{1}{4} \sum_{jkl} P_{ik}P_{lj}c_{ik,lj}(3x_{ij} - x_{il} - x_{kl} - x_{kj})$$

$$= \frac{1}{2} \sum_{jkl} P_{ik}P_{lj}c_{ik,lj}(x_{ij} - x_{kj})$$

$$= \frac{1}{2} \sum_{jkl} P_{ik}P_{lj}(s_{ij} - s_{kj}) + O(s^2). \qquad (4.82)$$

Assuming there has been at least one generation of random mating, (4.81) reduces (4.82) to

$$\Delta p_i = \frac{1}{2}p_i\,(s_i - \bar{s}) + O(s^2) \qquad (4.83)$$

where

$$s_i = \sum_j s_{ij}p_j, \qquad \bar{s} = \sum_{ij} s_{ij}p_ip_j.$$

For the same set of gene frequencies and selection coefficients, the lowest order term in (4.83) is just one-half of the corresponding result for the standard model, (4.25). Wright (1955) derived this result for two alleles by assuming the population was in Hardy-Weinberg proportions, a procedure formally justified by (4.81). Since a population will evolve quite differently under selection within the entire population and selection only within families, therefore, even with the same selection coefficients, the gene frequencies in the two cases will differ. Consequently, the rate of gene frequency change in the second case will not be one-half that in the first.

For two alleles, with the notation $P = P_{11}$, $Q = P_{12}$, $R = P_{22}$, the exact equations (4.81) become (King, 1965)

$$P' = P^2 + 4x_{11}Q\left(\frac{P}{x_{11}+x_{12}} + \frac{Q}{x_{11}+2x_{12}+x_{22}}\right), \qquad (4.84a)$$

$$Q' = PR + 2x_{12}Q\left(\frac{P}{x_{11}+x_{12}} + \frac{Q}{x_{11}+2x_{12}+x_{22}} + \frac{R}{x_{12}+x_{22}}\right), \qquad (4.84b)$$

$$R' = R^2 + 4x_{22}Q\left(\frac{Q}{x_{11}+2x_{12}+x_{22}} + \frac{R}{x_{12}+x_{22}}\right). \qquad (4.84c)$$

From (4.84) we find the gene frequency change

$$\Delta p = Q\left[P\left(\frac{x_{11}-x_{12}}{x_{11}+x_{12}}\right) + 2Q\left(\frac{x_{11}-x_{22}}{x_{11}+2x_{12}+x_{22}}\right) + R\left(\frac{x_{12}-x_{22}}{x_{12}+x_{22}}\right)\right]. \qquad (4.85)$$

For constant x_{ij}, we conclude from (4.85) at a glance that (as long as there is selection) $x_{11} \geq x_{12} \geq x_{22}$ implies $\Delta p > 0$ for $0 < p < 1$, so $p(t) \to 1$, as for the usual model. Similarly, if $x_{11} \leq x_{12} \leq x_{22}$, then $p(t) \to 0$. It is easy to check, however, that the Hardy-Weinberg over- or underdominant equilibrium (4.29) never satisfies (4.84). We shall show, nevertheless, that overdominance is sufficient for a protected polymorphism.

Consider $p \to 0$. Since $p = P + Q \geq P, Q$, we infer from (4.84a) that $P = O(p^2)$ after one generation. Therefore, $p = Q + O(p^2)$, and (4.85) becomes

$$\Delta p = p\left(\frac{x_{12}-x_{22}}{x_{12}+x_{22}}\right) + O(p^2),$$

whence we require $x_{12} > x_{22}$. Obviously, a will be protected if $x_{12} > x_{22}$, proving our assertion.

4.7 *Maternal Inheritance*

Characters are frequently affected by the maternal genotype. We shall discuss the extreme case of traits depending entirely on the mother's genotype. For examples, consult Srb, Owen, and Edgar (1965, Ch. 11). Let y_{ij} be the viability of an individual whose mother is A_iA_j. We postulate random mating. In terms of the frequency, P_{ij}, of A_iA_j adults, our difference equations read

$$P'_{ij} = (2\bar{y})^{-1} \sum_{kl} P_{ik}P_{lj}(y_{ik} + y_{lj}) \qquad (4.86)$$

$$= (2\bar{y})^{-1} p_i p_j (y_i + y_j), \qquad (4.87)$$

where

$$p_i y_i = \sum_j P_{ij} y_{ij}, \qquad \bar{y} = \sum_{ij} P_{ij} y_{ij}. \qquad (4.88)$$

We set $y_{ij} = 1 + s_{ij}$, and infer immediately from (4.87)

$$P'_{ij} = p_i p_j + O(s). \qquad (4.89)$$

Thus, as for intra-family selection, the deviation of the adults from Hardy-Weinberg proportions is reduced to the order of the selection intensity in one generation of random mating. Owing to the parental effect, the fact that zygotes are in Hardy-Weinberg ratios is again useless. From (4.87) we obtain

$$\Delta p_i = (2\bar{y})^{-1} p_i (y_i - \bar{y}), \tag{4.90}$$

which, for the same fitnesses and genotypic frequencies, is one-half the gene frequency change in the standard model. For two alleles, Wright (1969) made this observation in the Hardy-Weinberg approximation. The comments below (4.83) apply here.

With two alleles, (4.86) reduces to (Wright, 1969)

$$P' = \bar{y}^{-1} p(Py_{11} + Qy_{12}), \tag{4.91a}$$

$$Q' = (2\bar{y})^{-1}(qPy_{11} + Qy_{12} + pRy_{22}), \tag{4.91b}$$

$$R' = \bar{y}^{-1} q(Qy_{12} + Ry_{22}), \tag{4.91c}$$

where $P = P_{11}$, $Q = P_{12}$, $R = P_{22}$. From (4.90) we find

$$\Delta p = (2\bar{y})^{-1}[qP(y_{11}-y_{12}) + pR(y_{12}-y_{22})]. \tag{4.92}$$

For constant y_{ij}, it follows readily that $y_{11} \geq y_{12} \geq y_{22}$ implies $p(t) \to 1$, and $y_{11} \leq y_{12} \leq y_{22}$ implies $p(t) \to 0$. It is easily verified that the over- or underdominant equilibrium $\hat{P} = \hat{p}^2$, $\hat{Q} = \hat{p}\hat{q}$, $\hat{R} = \hat{q}^2$,

$$\hat{p} = \frac{y_{22} - y_{12}}{y_{11} - 2y_{12} + y_{22}},$$

satisfies (4.91). As in Section 4.6, we see that $P' = O(p^2)$ as $p \to 0$. Therefore, (4.92) informs us that overdominance, $y_{12} > y_{11}, y_{22}$, is sufficient for a protected polymorphism.

4.8 *Meiotic Drive*

Meiotic drive refers to non-Mendelian segregation at meiosis. Its effect is manifestly the same as that of selection acting independently among the gametes produced by each individual. The most thoroughly investigated examples are in *Drosophila* and the house mouse. Non-Mendelian segregation can maintain a deleterious allele in the population. Indeed, the "driven" allele is often lethal in the homozygous state. We shall show only how to formulate the equations describing meiotic drive. The analysis of realistic models requires considerable algebra (Hartl, 1970, 1970a).

Let α_{ij} be the probability that a successful gamete from an A_iA_j male carries A_i (Hartl, 1970b). Thus, $\alpha_{ji} = 1-\alpha_{ij}$, whence $\alpha_{ii} = 1/2$. For Mendelian segregation, $\alpha_{ij} = 1/2$ for all i,j. In most cases, the amount of segregation distortion is different in the two sexes. We denote the female segregation parameter corresponding to α_{ij} by β_{ij}. Let p_i and q_i, and x_{ij} and y_{ij} represent the frequencies of A_i in sperm and eggs, and the viabilities of A_iA_j males and females. Assuming random mating, the zygotic frequencies in both sexes are $\frac{1}{2}(p_iq_j + p_jq_i)$. Therefore,

$$p_i' = \bar{x}^{-1} \sum_j (p_iq_j + p_jq_i)x_{ij}\alpha_{ij},$$

$$q_i' = \bar{y}^{-1} \sum_j (p_iq_j + p_jq_i)y_{ij}\beta_{ij},$$

where

$$\bar{x} = \sum_{ij} (p_iq_j + p_jq_i)x_{ij}\alpha_{ij}$$

$$= \frac{1}{2} \sum_{ij} (p_iq_j + p_jq_i)x_{ij}(\alpha_{ij} + \alpha_{ji})$$

$$= \frac{1}{2} \sum_{ij} (p_iq_j + p_jq_i)x_{ij},$$

$$\bar{y} = \frac{1}{2} \sum_{ij} (p_iq_j + p_jq_i)y_{ij}.$$

Note that we require equal viabilities, $x_{ij} = y_{ij}$, and equal segregation parameters, $\alpha_{ij} = \beta_{ij}$, in order to have the same allelic frequencies in the two sexes, $p_i' = q_i'$.

4.9 *Mutation and Selection*

To introduce mutation, we follow the procedure of Section 2.2. Let P_{ij} be the frequency of A_iA_j zygotes. We denote the zygotic proportions in the next generation without taking mutation into account by P_{ij}^*. From (4.6) we have

$$P_{ij}^* = \bar{f}^{-1} \sum_{kl} X_{ik,lj}f_{ik,lj}, \tag{4.93}$$

where $\bar{f}$ is given by (4.7). While selection acts on the phenotype, which develops from the zygotic genotype, the germ cells mutate with no phenotypic effect. Hence, designating by R_{ij} the probability that an A_i allele in a zygote appears as A_j in a gamete, the zygotic frequencies in the next generation are

$$P_{ij}' = \sum_{kl} P_{kl}^*R_{ki}R_{lj}. \tag{4.94}$$

Equations (4.93) and (4.94) constitute our model.

The allelic frequencies satisfy much simpler equations. From (4.17), we see that

$$p_i^* = p_i w_i / \overline{w}, \tag{4.95}$$

the fitnesses being specified by (4.13) and (4.14). Since

$$\sum_j R_{ij} = 1,$$

(4.94) yields

$$p_i' = \sum_k p_k^* R_{ki}. \tag{4.96}$$

Denoting the mutation rates by u_{ij} ($u_{ii} = 0$) and substituting (2.26) into (4.96), we find

$$p_i' = p_i^*(1 - \sum_k u_{ik}) + \sum_k p_k^* u_{ki}. \tag{4.97}$$

If mating is random and fertilities are multiplicative, (4.24) informs us that

$$P_{ij}^* = p_i^* p_j^*. \tag{4.98}$$

Inserting (4.98) into (4.94) and employing (4.96) gives

$$P_{ij}' = p_i' p_j'. \tag{4.99}$$

Thus, (4.25b), (4.95), and (4.96) determine the evolution of the population in this case.

With weak selection and mutation, the argument in Section 2.2 again shows that the changes due to selection and mutation are approximately additive.

Assuming only that mutation is much weaker than selection, we shall relate the total frequency of mutants at equilibrium to the total mutation rate and the mean selection coefficient. This was done in continuous time by Crow and Kimura (1970). Let A_1 be the normal allele, A_i being deleterious for $i > 1$. Since mutation is much weaker than selection, $p_i << p_1$, $i > 1$. Then we may neglect the last term in (4.97) for $i = 1$. Without reverse mutation, (4.97) yields

$$p_1' = p_1^*(1 - u), \tag{4.100}$$

where

$$u = \sum_k u_{1k} \tag{4.101}$$

is the total forward mutation rate. Substituting (4.95) into (4.100), at equilibrium we find

$$\bar{w} = w_1(1-u). \tag{4.102}$$

We must approximate the fitnesses in (4.102). There are two distinct cases.

1. A_1 Completely Dominant

Choosing $w_{11} = 1$, we have $w_{1i} = 1$ for all i. We write $w_{ij} = 1-s_{ij}$, $s_{1i} = 0$, $s_{ij} \geq 0$, for all i,j. It suffices to suppose $u_{1i} \ll s_{ii}$ for all i. The total frequency of mutant phenotypes is

$$Q = \sum_{i,j>1} P_{ij}. \tag{4.103}$$

From (4.13) we find $w_1 = 1$, while from (4.14) we obtain

$$\bar{w} = \sum_{ij} P_{ij}(1-s_{ij}) = 1-Q\bar{s}, \tag{4.104}$$

where

$$\bar{s} = Q^{-1} \sum_{ij} P_{ij}s_{ij} \tag{4.105}$$

is the mean selection coefficient of the mutant phenotypes. Substituting our results into (4.102), we conclude

$$Q = u / \bar{s}, \tag{4.106}$$

i.e., the total frequency of mutant phenotypes is the ratio of the total forward mutation rate to the average selection coefficient.

This formula is approximate only because we ignored reverse mutation. For two alleles, recalling (4.101), (4.103), and (4.105), we obtain from (4.106)

$$P_{22} = u_{12} / s_{22}; \tag{4.107}$$

in Hardy-Weinberg proportions this reduces to

$$p_2 = (u_{12} / s_{22})^{1/2}. \tag{4.108}$$

2. A_1 Incompletely Dominant

We take again $w_{ij} = 1-s_{ij}$, $s_{11} = 0$, $s_{1i} > 0$, for all i,j. We assume $u_{1i} \ll s_{1i}$ for all i. From (4.13) we derive

$$p_1 w_1 = \sum_i P_{1i}(1-s_{1i}) = p_1 - T\tilde{s}, \tag{4.109}$$

where

$$T = \sum_{i>1} P_{1i} = p_1 - P_{11} \tag{4.110}$$

and

$$\tilde{s} = T^{-1} \sum_i P_{1i} s_{1i} \qquad (4.111)$$

are one-half the total frequency and the average selection coefficient of normal-mutant heterozygotes. The mean fitness reads

$$\bar{w} = \sum_{ij} P_{ij}(1 - s_{ij})$$

$$= 1 - 2 \sum_i P_{1i} s_{1i} - \sum_{i,j>1} P_{ij} s_{ij}$$

$$\approx 1 - 2T\tilde{s}. \qquad (4.112)$$

Since $p_i << 1$, we shall normally have $P_{ij} << P_{1i}$, $i,j > 1$. As we see from (4.60), however, this will fail with substantial inbreeding. Thus, unless A_1 is almost completely dominant, $s_{1i} << s_{ii}$, (4.112) will be a good approximation. Substituting (4.109) and (4.112) into (4.102) leads to

$$u \approx T\tilde{s}[2 - p_1^{-1}(1-u)] \approx T\tilde{s},$$

whence

$$T \approx u / \tilde{s}. \qquad (4.113)$$

But

$$p_i = P_{i1} + \sum_{j>1} P_{ij} \approx P_{1i}, \qquad (4.114)$$

so the sum in (4.110) yields

$$T \approx \sum_{i>1} p_i = 1 - p_1. \qquad (4.115)$$

Combining this with (4.113), we obtain the desired result,

$$1 - p_1 \approx u / \tilde{s}, \qquad (4.116)$$

where (4.111), (4.114), and (4.115) give

$$\tilde{s} \approx (1 - p_1)^{-1} \sum_i p_i s_{1i}. \qquad (4.117)$$

Thus, the total frequency of mutant alleles is approximately equal to the ratio of the total forward mutation rate to the mean normal-mutant heterozygote selection coefficient. For two alleles, (4.116) reduces to

$$p_2 \approx u_{12} / s_{12}, \qquad (4.118)$$

which, for comparable parameters, is a much lower frequency than the corresponding one, (4.108), for a recessive mutant. This is hardly surprising: selection against very rare recessive homozygotes is much less effective than selection against heterozygotes.

4.10 *Continuous Model with Overlapping Generations*

In this section, we shall construct a model of a continuously evolving population with age structure. It will be possible to carry over some of the results of Section 2.4 with only obvious modifications. With some simplifying assumptions, we shall reduce the general model to one involving only the total numbers of the various genotypes. We shall investigate the dynamics of the simplified model with particular attention to weak selection and the Fundamental Theorem of Natural Selection.

Our formulation of the age-structured model will be based on that of Cornette (1975). See also Norton (1928) and Charlesworth (1970, 1974). The notation will be, as far as possible, the diploid version of the notation of Section 2.4. Let $\nu_{ij}(t,x)\Delta x$ be the number of individuals between the ages of x and $x+\Delta x$ at time t. Then the total number of A_iA_j individuals and the total population size at time t are

$$n_{ij}(t) = \int_0^\infty \nu_{ij}(t,x)dx \tag{4.119}$$

and

$$N(t) = \sum_{ij} n_{ij}(t). \tag{4.120}$$

As in Section 2.4, we start observing the population at $t = 0$. For $t \geq x$, the life table, $l_{ij}(t,x)$ represents the probability that an A_iA_j individual born at time $t-x$ survives to age x. If $t < x$, $l_{ij}(t,x)$ designates the probability that an A_iA_j individual aged $x-t$ at time 0 survives to age x. Therefore,

$$l_{ij}(t,0) = 1 \quad \text{and} \quad l_{ij}(0,x) = 1.$$

Since the analysis and discussion between (2.46) and (2.54) do not depend on the mode of reproduction, they apply equally well to a diploid population. Thus, as in (2.47), the age distribution, $\nu_{ij}(t,x)$, is determined by the initial age distribution, $\nu_{ij}(0,x)$; life table, $l_{ij}(t,x)$; and the rate of birth of A_iA_j individuals per unit time, $\nu_{ij}(t,0)$:

$$\nu_{ij}(t,x) = \begin{cases} \nu_{ij}(t-x,0)l_{ij}(t,x), & t \geq x, \\ \nu_{ij}(0,x-t)l_{ij}(t,x), & t < x. \end{cases}$$

If $d_{ij}(t,x)\Delta x$ is the probability that an A_iA_j individual aged x at time t dies between ages x and $x+\Delta x$, from (2.49) we have

$$\frac{\partial l_{ij}}{\partial t} + \frac{\partial l_{ij}}{\partial x} = -d_{ij}(t,x)l_{ij}(t,x),$$

with the solution (see 2.52)

$$l_{ij}(t,x) = \begin{cases} \exp\left[-\int_0^x d_{ij}(\xi+t-x,\xi)d\xi \right], & t \geq x, \\ \exp\left[-\int_{x-t}^x d_{ij}(\xi+t-x,\xi)d\xi \right], & t < x. \end{cases}$$

We shall not write the obvious analogues of (2.58) and (2.59). Finally, (2.53) and (2.54) yield

$$\frac{\partial \nu_{ij}}{\partial t} + \frac{\partial \nu_{ij}}{\partial x} = -d_{ij}(t,x)\nu_{ij}(t,x)$$

and

$$\dot{n}_{ij}(t) = \nu_{ij}(t,0) - \int_0^\infty d_{ij}(t,x)\nu_{ij}(t,x)dx. \tag{4.121}$$

To reduce the birth term, $\nu_{ij}(t,0)$, in (4.121), we must consider the mating structure of the population. We designate by $Y_{ij,kl}(t,x,y)\Delta t\Delta x\Delta y$ the number of matings from time t to $t+\Delta t$ between A_iA_j individuals aged x to $x+\Delta x$ and A_kA_l individuals aged y to $y+\Delta y$. Let $a_{ij,kl}(t,x,y)$ be the average number of progeny from one such union. Then

$$\nu_{ij}(t,0) = \sum_{kl} \int_0^\infty dx \int_0^\infty dy\, Y_{ik,lj}(t,x,y)a_{ik,lj}(t,x,y). \tag{4.122}$$

We wish to impose on the mating frequencies, fertilities, and mortalities restrictions which will lead to a closed system of differential equations for $n_{ij}(t)$. As in Section 2.4, we shall do this two ways. First, we assume age-independence. For the fertilities and mortalities, this means

$$a_{ij,kl}(t,x,y) = a_{ij,kl}(t), \qquad d_{ij}(t,x) = d_{ij}(t). \tag{4.123}$$

We cannot suppose $Y_{ij,kl}(t,x,y) = Y_{ij,kl}(t)$ because $Y_{ij,kl}(t,x,y) = 0$ if $\nu_{ij}(t,x) = 0$ or $\nu_{kl}(t,y) = 0$. Therefore, we hypothesize

$$Y_{ij,kl}(t,x,y) = \overline{Y}_{ij,kl}(t)\nu_{ij}(t,x)\nu_{kl}(t,y). \tag{4.124}$$

Let $M(t)$ be the total number of matings per unit time and $X_{ij,kl}(t)$ the fraction of these matings between A_iA_j and A_kA_l. Thus, from (4.119) and (4.124) we obtain

$$M(t)X_{ij,kl}(t) = \int_0^\infty dx \int_0^\infty dy\ \overline{Y}_{ij,kl}(t)\nu_{ij}(t,x)\nu_{kl}(t,y)$$

$$= \overline{Y}_{ij,kl}(t)n_{ij}(t)n_{kl}(t), \tag{4.125}$$

showing that with the *Ansatz* (4.124) $X_{ij,kl}(t)$ is independent of age structure. Substituting (4.119), (4.123), (4.124), and (4.125) into (4.121) and (4.122) yields

$$\dot{n}_{ij}(t) = M(t)\sum_{kl} X_{ik,lj}(t)a_{ik,lj}(t) - d_{ij}(t)n_{ij}(t). \tag{4.126}$$

The dependence on time in our basic equation (4.126) includes possible dependence on the genotypic numbers, $\underset{\sim}{n}(t)$.

Let us derive (4.126) by assuming the population has reached a stable age distribution. We posit that the proportion of A_iA_j individuals in the age range x to $x+\Delta x$ is $\lambda_{ij}(x)\Delta x$, i.e., $\nu_{ij}(t,x) = n_{ij}(t)\lambda_{ij}(x)$. The hypothesis is that $\lambda_{ij}(x)$ is a known function of age only. Since

$$\int_0^\infty \lambda_{ij}(x)dx = 1,$$

the average death rate of A_iA_j individuals reads

$$d_{ij}(t) = \int_0^\infty d_{ij}(t,x)\lambda_{ij}(x)dx. \tag{4.127}$$

We suppose that the age- and time-dependence of the mating frequencies are uncoupled:

$$Y_{ij,kl}(t,x,y) = M(t)X_{ij,kl}(t)R_{ij,kl}(x,y), \tag{4.128}$$

where

$$\int_0^\infty dx \int_0^\infty dy\ R_{ij,kl}(x,y) = 1.$$

Integrating (4.128) over x and y justifies again interpreting $X_{ij,kl}(t)$ as the proportion of matings, regardless of age, between A_iA_j and A_kA_l. Substituting the results in this paragraph into (4.121) and (4.122) yields (4.126) with

$$a_{ij,kl}(t) = \int_0^\infty dx \int_0^\infty dy\ R_{ij,kl}(x,y)a_{ij,kl}(t,x,y), \tag{4.129}$$

the analogue of (4.127). Clearly, $a_{ij,kl}(t)$ is the average fertility of an $A_iA_j \times A_kA_l$ union.

The discrete model formulated in Section 4.1 applies exactly to some organisms, such as annual plants and some insects. Our discussion of the conditions required for the validity of the continuous model (4.126) suggests that this model may describe some populations of higher organisms more accurately than the discrete one.

We proceed now to derive from (4.126) equations for the genotypic and gene frequencies, P_{ij} and p_i. Our treatment follows Nagylaki and Crow (1974), who present also analyses of some special cases and generalizations to two loci and dioecious populations. It is convenient to use fertilities, $f_{ij,kl}$, relative to the number of individuals rather than the number of matings. With

$$f_{ij,kl}(t) = M(t)a_{ij,kl}(t) / N(t), \tag{4.130}$$

one expects $f_{ij,kl}(t)$ to be constant whenever $a_{ij,kl}(t)$ is. Inserting (4.130) into (4.126), we get

$$\dot{n}_{ij} = N \sum_{kl} X_{ik,lj} f_{ik,lj} - d_{ij} n_{ij}. \tag{4.131}$$

The mating frequencies are normalized,

$$\sum_{ijkl} X_{ij,kl} = 1,$$

and satisfy the obvious symmetries

$$X_{ij,kl} = X_{kl,ij} = X_{ji,kl}.$$

The fertilities possess the same symmetries. Therefore, (4.131) informs us that $n_{ij}(0) = n_{ji}(0)$ implies $\dot{n}_{ij}(0) = \dot{n}_{ji}(0)$, whence $n_{ij}(t) = n_{ji}(t)$, $t \geq 0$, as is necessary.

Recalling (4.120), we deduce from (4.131)

$$\dot{N} = \overline{m}N, \tag{4.132}$$

where the genotypic frequencies are

$$P_{ij} = n_{ij} / N, \tag{4.133}$$

and the mean mortality, fertility, and fitness read

$$\overline{d} = \sum_{ij} P_{ij} d_{ij}, \tag{4.134}$$

$$\overline{f} = \sum_{ijkl} X_{ij,kl} f_{ij,kl}, \tag{4.135}$$

$$\bar{m} = \bar{f} - \bar{d}. \tag{4.136}$$

Differentiating (4.133) and substituting (4.131) and (4.132) gives

$$\dot{P}_{ij} = \sum_{kl} X_{ik,lj} f_{ik,lj} - (d_{ij} + \bar{m}) P_{ij}. \tag{4.137}$$

This is analogous to the basic equation (4.6) of the discrete model; like (4.6), it requires specification of the mating frequencies and the birth and death rates.

The number of offspring of a single $A_i A_j$ individual per unit time, b_{ij}, is given by

$$P_{ij} b_{ij} = \sum_{kl} X_{ij,kl} f_{ij,kl}. \tag{4.138}$$

The mean fecundity of the population is

$$\bar{b} = \sum_{ij} P_{ij} b_{ij} = \bar{f}. \tag{4.139}$$

To derive the differential equation satisfied by the allelic frequencies, p_i, we introduce allelic fertilities, mortalities, and fitnesses by the definitions

$$p_i b_i = \sum_j P_{ij} b_{ij}, \tag{4.140}$$

$$p_i d_i = \sum_j P_{ij} d_{ij}, \tag{4.141}$$

$$m_i = b_i - d_i. \tag{4.142}$$

Summing (4.137) over j and substituting (4.138), (4.140), (4.141), and (4.142), we obtain the fundamental equation for gene frequency change:

$$\dot{p}_i = p_i (m_i - \bar{m}). \tag{4.143}$$

This has the same form as (2.63), but the definition of the fitnesses is more complicated. Defining the Malthusian parameter of $A_i A_j$ as

$$m_{ij} = b_{ij} - d_{ij}, \tag{4.144}$$

we infer from (4.140), (4.141), (4.142),

$$p_i m_i = \sum_j P_{ij} m_{ij}, \tag{4.145}$$

and from (4.136), (4.139), (4.144),

$$\bar{m} = \sum_{ij} m_{ij} P_{ij}. \tag{4.146}$$

Thus, (4.143) involves the fertilities and mortalities explicitly only in the Malthusian parameter combination (4.144). This observation, however, is deceptive; the evolution of the genotypic proportions, controlled by (4.137), depends separately on the birth and death rates, and P_{ij} must be specified as a function of allelic frequencies in order to complete (4.143).

Henceforth, we shall suppose mating is random:

$$X_{ij,kl} = P_{ij}P_{kl}. \tag{4.147}$$

This simplifies (4.138) to

$$b_{ij} = \sum_{kl} P_{kl}f_{ij,kl}, \tag{4.148}$$

but does not lead to Hardy-Weinberg proportions. For zygotes to be in Hardy-Weinberg ratios, just as in the discrete case, we must restrict the fertilities. The correct condition in continuous time is additivity of the birth rates (Nagylaki and Crow, 1974). As intuitively expected, however, even if the population is originally in Hardy-Weinberg proportions, dominance in the mortalities produces deviations from Hardy-Weinberg ratios (Nagylaki and Crow, 1974). The variables

$$Q_{ij} = P_{ij} - p_i p_j \tag{4.149}$$

are suitable measures of these departures.

We shall now analyze (4.137) for weak selection (Nagylaki, 1976),

$$f_{ij,kl} = b + O(s), \qquad d_{ij} = d + O(s). \tag{4.150}$$

Genotype-independent mortality clearly has no effect on genotypic proportions; indeed, d cancels out of (4.137). We assume the fertility b is constant. At this stage, the selection pattern is completely arbitrary, and may depend on time and genotypic frequencies. The positive constant s gives the order of magnitude of the intensity of selection. If the selective differences are constant, we can always take s to be the largest selection coefficient. We are interested in the most common biological situation, $s \ll b$. Since the population size will usually be approximately stabilized, if time is measured in generations, we shall have $b \approx 1$, whence $s \ll 1$.

Substituting (4.147) and (4.150) into (4.137) yields

$$\dot{P}_{ij} = -bQ_{ij} + O(s), \tag{4.151}$$

while (4.143) and (4.150) tell us

$$\dot{p}_i = O(s). \tag{4.152}$$

From (4.149), (4.151), and (4.152) we conclude

$$\dot{Q}_{ij} = -bQ_{ij} + sg_{ij}(\underset{\sim}{P},t), \tag{4.153}$$

where $g_{ij}(\underset{\sim}{P},t)$ is a complicated function of birth and death rates, genotypic frequencies, and possibly of time, of order unity. More precisely, we may assume $g_{ij}(\underset{\sim}{P},t)$ is uniformly bounded for $t \geq 0$ as $s \to 0$. Our analysis will be based on (4.152) and (4.153). Observe that in the absence of selection ($s = 0$), $Q_{ij}(t)$ is reduced to zero at the rapid exponential rate e^{-bt}. We shall exploit this behavior for weak selection.

By the variation-of-parameters formula (see, e.g., Brauer and Nohel, 1969, p.72), (4.153) has the unique solution

$$Q_{ij}(t) = Q_{ij}(0)e^{-bt} + se^{-bt}\int_0^t e^{b\tau}g_{ij}[\underset{\sim}{P}(\tau),\tau]dt. \tag{4.154}$$

We choose t_1 as the shortest time such that

$$|Q_{ij}(0)|e^{-bt_1} \leq s \tag{4.155}$$

for all i,j. Of course, if the system starts sufficiently close to Hardy-Weinberg proportions, t_1 will be zero. Otherwise, we infer from (4.155) that roughly $t_1 \approx -b^{-1}\ln s$, which will usually be about 5 or 10 generations. Since (4.154) and (4.155) yield $Q_{ij}(t) = O(s)$ for $t \geq t_1$, we write $Q_{ij}(t) = sQ^0_{ij}(t)$. Thus, the population approaches the Hardy-Weinberg surface, $Q_{ij} = 0$, very rapidly, but at $t = t_1$ it is still very far from gene frequency change equilibrium, the gene frequency change from $t = 0$ to $t = t_1$ being quite small (crudely, $st_1 \approx -b^{-1}s\ln s$). For $t \geq t_1$, (4.153) reads

$$\dot{Q}^0_{ij} = -bQ^0_{ij} + g_{ij}(\underset{\sim}{P},t). \tag{4.156}$$

From (4.149), (4.152), and (4.156) we conclude that $\dot{P}_{ij} = O(s)$ for $t \geq t_1$. Assuming

$$\frac{\partial g_{ij}}{\partial t} = O(s), \tag{4.157}$$

(4.156) yields

$$\frac{d}{dt}[Q^0_{ij} - b^{-1}g_{ij}(\underset{\sim}{P},t)] = -b[Q^0_{ij} - b^{-1}g_{ij}(\underset{\sim}{P},t)] + O(s). \tag{4.158}$$

Recalling (4.153), we note that (4.157) requires that the explicit dependence of the selection coefficients on time, if any, be of $O(s^2)$. Arbitrary dependence on the genotypic frequencies is permitted. Writing the analogue of (4.154) for the bracket in (4.158), we obtain

$$Q^0_{ij}(t)-b^{-1}g_{ij}[\underset{\sim}{P}(t),t] = \{Q^0_{ij}(t_1)-b^{-1}g_{ij}[\underset{\sim}{P}(t_1),t_1]\}e^{-b(t-t_1)}+O(s). \tag{4.159}$$

Therefore, we define t_2 ($\geq t_1$) as the shortest time such that

$$|Q^0_{ij}(t_1) - b^{-1}g_{ij}[\underset{\sim}{P}(t_1), t_1]|e^{-b(t_2 - t_1)} \leq s. \qquad (4.160)$$

Since the coefficient of the exponential in (4.160) is of order unity, roughly, $t_2 - t_1 \approx t_1$ or $t_2 \approx 2t_1$. From (4.159) we find for $t \geq t_2$

$$Q^0_{ij}(t) = b^{-1}g_{ij}[\underset{\sim}{P}(t), t] + O(s). \qquad (4.161)$$

But $\dot{P}_{ij}(t) = O(s)$ for $t \geq t_1$. Hence, with the aid of (4.157), we deduce from (4.161) that $\dot{Q}^0_{ij}(t) = O(s)$ or

$$\dot{Q}_{ij}(t) = O(s^2), \qquad t \geq t_2, \qquad (4.162)$$

which is the desired result.

Notice that the system is not close to equilibrium until a time $t_3 \approx s^{-1} >> t_2$ has elapsed, so (4.162) cannot be obtained by local analysis. During the period $t_1 \leq t < t_2$, $\dot{p}_i = O(s)$ and $\dot{Q}_{ij} = O(s)$. Consequently, these quantities change by only about $s(t_2 - t_1) \approx -b^{-1}s \ln s$ during this epoch. For $t_2 \leq t < t_3$, by (4.162), Q_{ij} changes only by roughly $s^2(t_3 - t_2) \approx s$, while p_i approaches equilibrium closely.

Our results are easily displayed diagrammatically in the two-allele case. We put $p = p_1$, $q = p_2$, $Q = Q_{12}$, and use Q,p as coordinates (see Problem 3.2). It is trivial to see that $Q_{11} = Q_{22} = -Q$. The constraints $P_{11} \geq 0$, $P_{22} \geq 0$, $P_{11} + P_{22} \leq 1$ imply that the population evolves in the region bounded by the three parabolas in Figure 4.1. Hardy-Weinberg proportions correspond to the dashed line $Q = 0$, $0 \leq p \leq 1$. The times t_1, t_2, and t_3 are indicated on a typical trajectory, which is nearly horizontal for $0 \leq t < t_1$ and close to vertical for $t \geq t_2$. The system moves much faster on the "horizontal" than on the "vertical" part of the trajectory, though the rate of gene frequency change is always $\dot{p} = O(s)$. The equilibrium satisfies $Q(\infty) = O(s)$.

Before applying (4.162) to the evolution of a character controlled by a single locus, we shall demonstrate that the dynamics of the genotypic frequencies can be approximated with an error of $O(s)$ by the behavior of the much simpler system on the Hardy-Weinberg surface. Consider the population at time t_1. The point on the trajectory at that time corresponds to the set of gene frequencies $p_i(t_1)$, which evolve according to the complicated law, $p_i(t)$, calculated from (4.137) and (4.147). On the Hardy-Weinberg surface, there is a unique point with gene frequencies $\pi_i(t_1) = p_i(t_1)$, and these evolve according to the much simpler law, $\pi_i(t)$, calculated by imposing Hardy-Weinberg proportions on (4.143). In Figure 4.1, of course, $\pi(t_1) = p(t_1)$, a

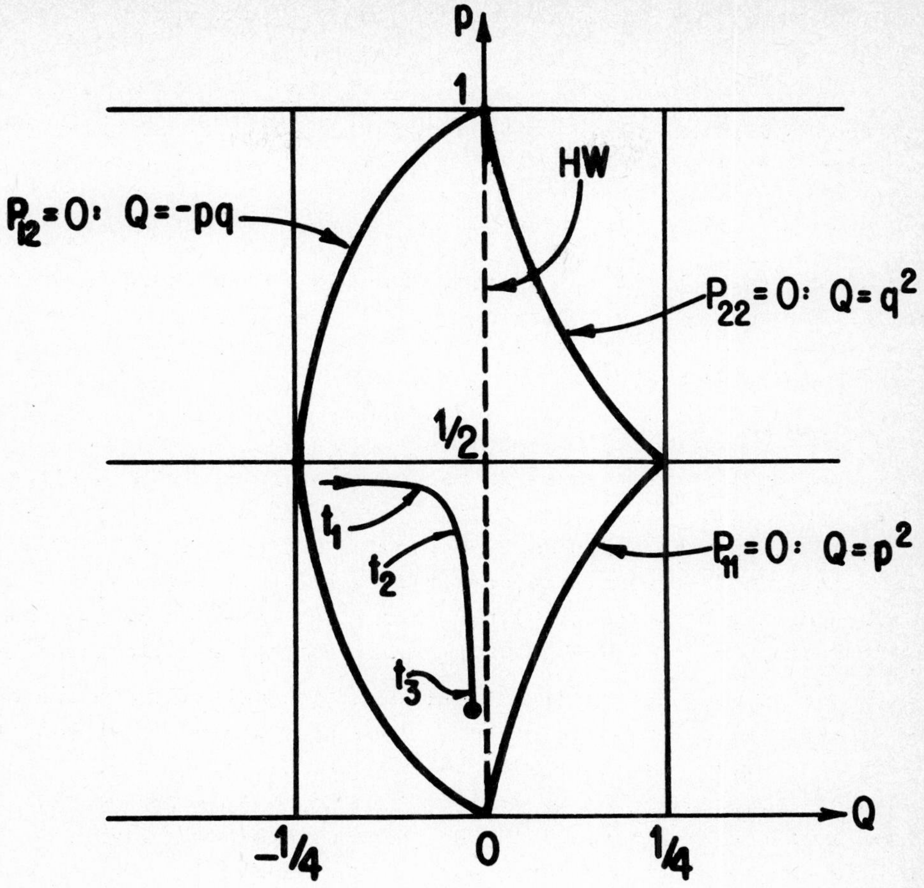

Fig. 4.1. The evolution of a typical diallelic single-locus system.

point on the p-axis found by horizontal projection. We shall show that $P_{ij}(t) = \pi_i(t)\pi_j(t) + O(s)$ for $t \geq t_1$.

We display the possible dependence of the genotypic fitnesses on the genotypic frequencies and time, and set

$$m_{ij}(\underset{\sim}{P},t) = m(\underset{\sim}{P},t) + sr_{ij}(\underset{\sim}{P},t), \tag{4.163}$$

where $r_{ij} = O(1)$. From (4.145) and (4.146) we obtain

$$m_i = m + sr_i, \qquad \overline{m} = m + s\overline{r}, \tag{4.164}$$

in which

$$p_i r_i(\underset{\sim}{P},t) = \sum_j r_{ij}(\underset{\sim}{P},t)P_{ij}, \qquad \overline{r}(\underset{\sim}{P},t) = \sum_{ij} r_{ij}(\underset{\sim}{P},t)P_{ij}. \tag{4.165}$$

Substituting (4.164) into (4.143), we find

$$\dot{p}_i = sp_i[r_i(\underset{\sim}{P},t) - \overline{r}(\underset{\sim}{P},t)]. \tag{4.166}$$

Recalling (4.149), we see that

$$r_{ij}(\underset{\sim}{P},t) = r_{ij}(\underset{\sim}{p}*\underset{\sim}{p},t) + O(\underset{\sim}{Q}), \tag{4.167}$$

where the asterisk indicates evaluation at $P_{ij} = p_i p_j$. Inserting (4.149) and (4.167) into (4.165) produces

$$r_i(\underset{\sim}{P},t) = c_i(\underset{\sim}{p},t) + O(\underset{\sim}{Q}), \qquad \overline{r}(\underset{\sim}{P},t) = \overline{c}(\underset{\sim}{p},t) + O(\underset{\sim}{Q}), \tag{4.168}$$

where

$$c_i(\underset{\sim}{p},t) = \sum_j r_{ij}(\underset{\sim}{p}*\underset{\sim}{p},t)p_j, \qquad \overline{c}(\underset{\sim}{p},t) = \sum_i c_i(\underset{\sim}{p},t)p_i. \tag{4.169}$$

Substituting (4.168) into (4.166) gives

$$\dot{p}_i = sp_i[c_i(\underset{\sim}{p},t) - \overline{c}(\underset{\sim}{p},t)] + O(s\underset{\sim}{Q}). \tag{4.170}$$

Equation (4.170) shows that the allelic frequencies, $\pi_i(t)$, on the Hardy-Weinberg surface satisfy

$$\dot{\pi}_i = s\pi_i[c_i(\underset{\sim}{\pi},t) - \overline{c}(\underset{\sim}{\pi},t)]. \tag{4.171}$$

For $t \geq t_1$, we have proved that $Q_{ij} = O(s)$, which implies that the error in (4.170) is of $O(s^2)$. If the Hardy-Weinberg system (4.171) tends to isolated equilibrium points, it follows by comparing (4.170) and (4.171) that

$$p_i(t) = \pi_i(t) + O(s), \qquad t \geq t_1. \tag{4.172}$$

Combining (4.149) and (4.172), we deduce

$$P_{ij}(t) = \pi_i(t)\pi_j(t) + O(s), \qquad t \geq t_1, \tag{4.173}$$

as claimed. If, due to unusual frequency or time dependence, or degeneracy, the system does not converge to isolated equilibrium points, small perturbations may cause large deviations in its ultimate state. Then (4.173) still holds for $t_1 \leq t \leq T$, for any fixed time T.

Finally, let us examine the evolution of the mean value, $\overline{Z}$, of a single-locus trait. For the same calculation for two loci, the reader may consult Crow and Nagylaki (1976). We shall slightly generalize Kimura's (1958) analysis of fitness, which we shall discuss as a special case. If Z_{ij} designates the character value of A_iA_j,

$$\overline{Z} = \sum_{ij} Z_{ij}P_{ij}. \tag{4.174}$$

Differentiating (4.174) yields

$$\dot{\bar{Z}} = \bar{\dot{Z}} + \sum_{ij} Z_{ij}\dot{P}_{ij}, \tag{4.175}$$

where

$$\bar{\dot{Z}} = \sum_{ij} \dot{Z}_{ij}P_{ij}, \tag{4.176}$$

the mean rate of change of the genotypic values. Introducing the deviations from the mean, $z_{ij} = Z_{ij} - \bar{Z}$, and noting that

$$\sum_{ij} \dot{P}_{ij} = \frac{d}{dt} \sum_{ij} P_{ij} = 0,$$

we obtain from (4.175)

$$\dot{\bar{Z}} = \bar{\dot{Z}} + \sum_{ij} z_{ij}\dot{P}_{ij}. \tag{4.177}$$

It is convenient to measure the departure from Hardy-Weinberg proportions by $\theta_{ij} = P_{ij}/(p_i p_j)$. Substituting

$$\dot{P}_{ij} = \dot{p}_i p_j \theta_{ij} + p_i \dot{p}_j \theta_{ij} + p_i p_j \dot{\theta}_{ij}$$

into (4.177) leads to

$$\dot{\bar{Z}} = \bar{\dot{Z}} + \sum_{ij} (2\dot{p}_i p_j \theta_{ij} + p_i p_j \dot{\theta}_{ij}) z_{ij}. \tag{4.178}$$

We can reduce (4.178) by performing a least squares decomposition of the *total genetic variance*

$$V = \sum_{ij} P_{ij} z_{ij}^2. \tag{4.179}$$

We write

$$z_{ij} = \alpha_i + \alpha_j + \Delta_{ij}, \tag{4.180}$$

where α_i is the *average effect* of A_i on the character, and Δ_{ij}, which is zero in the absence of dominance, is the *dominance deviation*. The procedure requires minimizing the *dominance variance*,

$$V_d = \sum_{ij} P_{ij}\Delta_{ij}^2 = \sum_{ij} P_{ij}(z_{ij} - \alpha_i - \alpha_j)^2, \tag{4.181}$$

with respect to α_i. We have

$$-\frac{1}{4}\frac{\partial V_d}{\partial \alpha_i} = \sum_j P_{ij}(z_{ij} - \alpha_i - \alpha_j) = \sum_j P_{ij}\Delta_{ij} = 0. \tag{4.182}$$

Although not needed here, we shall show that the variances are additive. The *genic* (or *additive genetic*) variance is

$$V_g = \sum_{ij} P_{ij}(\alpha_i + \alpha_j)^2. \tag{4.183}$$

Inserting (4.180) into (4.179) and using (4.181) and (4.183), we find

$$V = V_g + V_d + 2 \sum_{ij} P_{ij}(\alpha_i + \alpha_j)\Delta_{ij}$$

$$= V_g + V_d, \tag{4.184}$$

the sum vanishing due to (4.182). Using (4.180) and (4.182), the *average excess*, z_i, of A_i can be expressed as

$$p_i z_i = \sum_j P_{ij} z_{ij} = p_i \alpha_i + \sum_j P_{ij} \alpha_j. \tag{4.185}$$

Observe that

$$\sum_i p_i z_i = \sum_{ij} P_{ij} z_{ij} = 0, \tag{4.186}$$

and

$$2 \sum_i p_i \alpha_i = \sum_{ij} P_{ij} z_{ij} = 0, \tag{4.187}$$

i.e., the means of the average effects and excesses are zero. Furthermore, from (4.183) and (4.185) we deduce

$$V_g = \sum_{ij} P_{ij}(\alpha_i + \alpha_j)(\alpha_i + \alpha_j)$$

$$= 2 \sum_{ij} P_{ij}(\alpha_i + \alpha_j)\alpha_i$$

$$= 2 \sum_i p_i z_i \alpha_i. \tag{4.188}$$

In Hardy-Weinberg proportions, (4.185) and (4.187) inform us that $z_i = \alpha_i$, so

$$V_g = 2 \sum_i p_i z_i^2. \tag{4.189}$$

Let us return to $\dot{\bar{Z}}$. We substitute $P_{ij} = p_i p_j \theta_{ij}$ into (4.185), and the result,

$$\sum_j p_j \theta_{ij} z_{ij} = \alpha_i + \sum_j p_j \theta_{ij} \alpha_j, \tag{4.190}$$

and (4.180) into (4.178):

$$\dot{\bar{Z}} = \bar{\dot{Z}} + 2 \sum_i \dot{p}_i(\alpha_i + \sum_j p_j \theta_{ij} \alpha_j) + \sum_{ij} p_i p_j \dot{\theta}_{ij}(\alpha_i + \alpha_j + \Delta_{ij})$$

$$= \bar{\dot{Z}} + 2\sum_i \dot{p}_i\alpha_i + 2\sum_{ij} p_j\alpha_j(\dot{p}_i\theta_{ij} + p_i\dot{\theta}_{ij}) + \sum_{ij} p_ip_j\theta_{ij}\overset{\circ}{\theta}_{ij}\Delta_{ij}. \quad (4.191)$$

In (4.191),

$$\overset{\circ}{\theta} = \frac{d}{dt}\ln\theta$$

is the logarithmic derivative. But

$$\sum_i p_i\theta_{ij} = \sum_i p_j^{-1}P_{ij} = 1.$$

Therefore,

$$0 = \frac{d}{dt}\sum_i p_i\theta_{ij} = \sum_i (\dot{p}_i\theta_{ij} + p_i\dot{\theta}_{ij}),$$

telling us that the third term in (4.191) is zero. Recalling (4.143), we arrive at the desired formula:

$$\dot{\bar{Z}} = \mathrm{Cov}\,(\mu,\alpha) + \bar{\dot{Z}} + \overline{\overset{\circ}{\theta}\Delta}, \quad (4.192)$$

where

$$\mathrm{Cov}\,(\mu,\alpha) = 2\sum_i p_i\mu_i\alpha_i, \quad (4.193)$$

$$\overline{\overset{\circ}{\theta}\Delta} = \sum_{ij} P_{ij}\overset{\circ}{\theta}_{ij}\Delta_{ij}, \quad (4.194)$$

and $\mu_i = m_i - \bar{m}$ is the average excess for fitness of A_i.

The covariance in (4.192) is the *genic covariance* of the average excess of fitness and the average effect on the character. The fitnesses appear explicitly only in this term; selection influences the other terms through the genotypic frequencies. With constant genotypic values, the second term in (4.192) is zero. The last term vanishes if either there is no dominance for the trait or the population is in Hardy-Weinberg proportions.

Since $\mu_i = O(s)$, the covariance in (4.192) is of $O(s)$. Now,

$$\theta_{ij} = 1 + Q_{ij}/(p_ip_j),$$

and for $t \geq t_2$, $\dot{Q}_{ij} = O(s^2)$, $\dot{p}_i = O(s)$, and $Q_{ij} = O(s)$. Hence, $\dot{\theta}_{ij} = O(s^2)$, and (4.192) has the form

$$\dot{\bar{Z}} = \mathrm{Cov}\,(\mu,\alpha) + \bar{\dot{Z}} + O(s^2), \qquad t \geq t_2. \quad (4.195)$$

If the genotypic values change very slowly, so that $\bar{\dot{Z}} = o(s)$, we obtain the approximation $\dot{\bar{Z}} \approx \mathrm{Cov}\,(\mu,\alpha)$. This may fail if the covariance is particularly small. Since $\mathrm{Cov}\,(\mu,\alpha) = 0$ at equilibrium because $\dot{p}_i = p_i\mu_i = 0$, the failure is most likely to occur near equilibrium.

If the trait is fitness, (4.188) and (4.193) give $\mathrm{Cov}(\mu,\alpha) = V_g$, the genic variance in fitness, and (4.192) becomes

$$\dot{\bar{m}} = V_g + \bar{\dot{m}} + \overline{\overset{\circ}{\theta}\Delta}, \tag{4.196}$$

the form of the Fundamental Theorem of Natural Selection due to Kimura (1958). Of course $V_g = O(s^2)$ and $\Delta_{ij} = O(s)$, whence

$$\dot{\bar{m}} = V_g + \bar{\dot{m}} + O(s^3), \qquad t \geq t_2. \tag{4.197}$$

With nearly constant fitnesses, i.e., $\bar{\dot{m}} = o(s^2)$, we get the approximation $\dot{\bar{m}} \approx V_g$. Again, this approximation may occasionally fail, particularly in the neighborhood of an equilibrium. Observe that $\dot{m}_{ij} = o(s^2)$ is a much stronger condition than the requirement (see 4.157)

$$\frac{\partial m_{ij}}{\partial t} = O(s^2),$$

involving only the explicit time dependence, which we imposed in the analysis. We conclude from the result $\dot{\bar{m}} \approx V_g$ that the mean fitness will generally increase for $t \geq t_2$.

For further results, the reader may refer to (Nagylaki, 1976, 1977a).

4.11 *Problems*

4.1. For the model of Section 4.1, show that if the zygotes are in Hardy-Weinberg proportions and viabilities are multiplicative, the adults will be in Hardy-Weinberg proportions. Express v_{ij} in terms of v_i.

4.2. Show that for multiplicative fitnesses (4.25) reduces to the haploid model, (2.6).

4.3. Assume $w_{11} = 1$, $w_{12} = w$, $w_{22} = 0$ in (4.26). Calculate $q(t)$ for constant w. Compare $q(\infty)$ and the rate of convergence to the results in Section 4.2.

4.4. The monotonicity property of the haploid model described in Problem 2.2 also holds for the diallelic diploid model, (4.26), with constant fitnesses. Prove this by deducing that

$$\frac{dp'}{dp} = \frac{qw_{22}w_1 + pw_{11}w_2}{\bar{w}^2} > 0.$$

4.5. Suppose the continuous mapping $p' = f(p)$ of $[a,b]$ into itself has a fixed point $\hat{p} = f(\hat{p})$ in (a,b). Define $x = p - \hat{p}$, and assume the mapping can be rewritten in the form $x' = \lambda(p)x$, with $-1 < \lambda(p) < 1$ for $a < p < b$. Suppose also that there is no p in (a,b) such that $f(p) = a$ or $f(p) = b$. Prove that $\hat{p}$ is unique and $p(t) \to \hat{p}$ as $t \to \infty$.

4.6. To see how to approximate geometric rates of convergence to the second order, consider the difference equation

$$x' = \alpha x(1 + \beta x), \qquad |\alpha| < 1.$$

Suppose $|\alpha|[1 + |\beta x(0)|] < 1$, which is certainly true for sufficiently small $|x(0)|$. Show that as $t \to \infty$

$$x(t) \approx c\alpha^t - \frac{\beta c^2 \alpha^{2t}}{1-\alpha},$$

where c is a constant.

4.7. For the fitness patterns displayed below, locate all the equilibria; state whether they are stable, asymptotically stable, or unstable; and, as far as possible with the results in Section 4.3, state where the system ultimately converges.

(a)
$$W = \begin{pmatrix} 1 & 2 & 3 \\ 2 & 2 & 1 \\ 3 & 1 & 1 \end{pmatrix},$$

(b)
$$W = \begin{pmatrix} 1 & 3 & 2 \\ 3 & 1 & 2 \\ 2 & 2 & 2 \end{pmatrix},$$

(c)
$$W = \begin{pmatrix} 1 & 2 & 2 \\ 2 & 1 & 2 \\ 2 & 2 & 1 \end{pmatrix}.$$

4.8. A_1 and A_2 are in stable polymorphic equilibrium. A small proportion of A_3 is introduced into the population. If the fitness pattern is

$$W = \begin{pmatrix} 1 & 3 & 1 \\ 3 & 2 & 4 \\ 1 & 4 & \cdot \end{pmatrix},$$

is A_3 established?

4.9. Consider the model (4.26) with frequency-dependent fitnesses $w_{ij}(p)$. Assuming that the heterozygote fitness, $w_{12}(p)$, cannot vanish simultaneously with either homozygote fitness, obtain sufficient conditions for a protected polymorphism if the allele A is

(i) neither completely dominant nor recessive,

(ii) completely dominant.

4.10. Prove (4.76).

4.11. In the diallelic, Hardy-Weinberg, mutation-selection model, take $w_{11} = w_{12} = 1$, $w_{22} = 1 - s$, $u_{12} = u$, $u_{21} = 0$, $0 < u \leq s \leq 1$. Prove that there is global nonoscillatory convergence to a unique equilibrium. What is the asymptotic rate of convergence? Show that for $u << 1$ this is approximately $\exp[-2(\sqrt{us} - u)t]$. Of course, this reduces to $\exp(-2\sqrt{us}t)$ in the usual case, $u << s$.

4.12. The genotypic frequencies for inbreeding with multiple alleles are

$$P_{ij} = p_i F \delta_{ij} + p_i p_j (1 - F).$$

Use (4.194) to show that if F is constant,

$$\overline{\overset{\circ}{\theta} \Delta} = -F \sum_i p_i \mu_i \Delta_{ii}.$$

5. NONRANDOM MATING

As discussed in Chapter 1, systems of nonrandom mating may depend on relatedness (inbreeding) or phenotype (assortative or disassortative mating). In the former case, the effect is the same on all loci; in the absence of selection, gene frequencies are conserved. Assortation or disassortation may be directly for the trait under investigation or for a character correlated with it due to common genetic or environmental factors. The effect on each locus, if any, depends on the genetic determination of the phenotype. In linkage equilibrium, loci not influencing the trait are unaffected. Since some genotypes may be more likely to mate than others, gene frequencies may change even if all matings have the same fertility and all genotypes are equally viable. Inbreeding increases homozygosity and genetic variability; to the extent that genotype and phenotype are correlated, assortative mating will have similar effects.

The variety of models and the diversity of their dynamical behavior are limitless (Karlin, 1968). We have already considered selection with a fixed inbreeding coefficient in Section 4.4. We shall analyze selfing with selection in Section 5.1. Our first assortative mating model is equivalent to partial self-fertilization without selection (Section 5.2). Assortative mating with complete dominance will be treated in Section 5.3. The model of Section 5.4 is best interpreted as random mating with differential (nonmultiplicative) fertility. Sections 5.5 and 5.6 are motivated by existing incompatibility systems in plants.

Before analyzing particular models, let us return to the important question of the possible gene frequency change due to differential probability of mating. In this chapter, we confine ourselves to a single autosomal locus, and assume that the sexes need not be distinguished. As usual, we write p_i, P_{ij}, and $X_{ij,kl}$ for the allelic, genotypic, and mating frequencies. If all matings are equally fertile and there are no viability differences, (4.8) informs us that the fertility of A_iA_j is given by

$$P_{ij}b_{ij} = \sum_{kl} X_{ij,kl}. \qquad (5.1)$$

The right-hand side of (5.1) is the total proportion of matings in which

A_iA_j participates as "first" partner. If this proportion is the same as the frequency of A_iA_j in the population, i.e., if

$$P_{ij} = \sum_{kl} X_{ij,kl}, \tag{5.2}$$

then we infer from (5.1) that $b_{ij} = 1$. If all genotypes have the same fertility, (4.17) tells us that the gene frequencies are constant: $p_i' = p_i$. Thus, (5.2) is sufficient for gene frequency conservation. The necessary condition

$$p_i = \sum_{jkl} X_{ik,lj} \tag{5.3}$$

follows directly from (4.6).

5.1 *Selfing with Selection*

The offspring from the self-fertilization of A_iA_j are the same as from the cross $A_iA_j \times A_iA_j$. We confine ourselves to two alleles, and use the numbers x_1, x_2, x_3 of AA, Aa, aa zygotes as our basic variables. Since the numbers satisfy linear equations, they are more convenient than the frequencies

$$y_i = x_i \Big/ \sum_j x_j. \tag{5.4}$$

With viabilities v_i, the adult numbers are v_ix_i. Therefore, with fertilities f_i and fitnesses $w_i = f_iv_i$, the recursion relations read

$$x_1' = w_1x_1 + \tfrac{1}{4}w_2x_2, \tag{5.5a}$$

$$x_2' = \tfrac{1}{2}w_2x_2, \tag{5.5b}$$

$$x_3' = \tfrac{1}{4}w_2x_2 + w_3x_3. \tag{5.5c}$$

We rewrite (5.5) in matrix form, and suppose henceforth that the fitnesses are constant:

$$\underset{\sim}{x}' = B\underset{\sim}{x}, \tag{5.6}$$

where

$$B = \begin{pmatrix} w_1 & \tfrac{1}{4}w_2 & 0 \\ 0 & \tfrac{1}{2}w_2 & 0 \\ 0 & \tfrac{1}{4}w_2 & w_3 \end{pmatrix}. \tag{5.7}$$

Since matrix difference equations of the form (5.6) are very common in population genetics, we shall explain how to analyze them in general. Regardless of the number of dimensions, n, and the form of B, (5.6) is immediately iterated to

$$\underset{\sim}{x}(t) = B^t \underset{\sim}{x}(0). \tag{5.8}$$

If B has a complete set of eigenvectors $\underset{\sim}{V}_i$, which it necessarily has if all the eigenvalues are distinct, we have

$$\underset{\sim}{x}(0) = \sum_{j=1}^{n} c_j \underset{\sim}{V}_j, \tag{5.9}$$

where the constants c_j are determined from the initial numbers, $\underset{\sim}{x}(0)$. Substituting (5.9) into (5.8), we obtain the desired result,

$$\underset{\sim}{x}(t) = \sum_{j=1}^{n} c_j \lambda_j^t \underset{\sim}{V}_j, \tag{5.10}$$

where the eigenvalues λ_j of B satisfy

$$B\underset{\sim}{V}_j = \lambda_j \underset{\sim}{V}_j. \tag{5.11}$$

Let λ_i, $i = 1, 2, \ldots, m \leq n$ be the eigenvalue(s) of greatest modulus, i.e., $\lambda_i = \mu\omega_i$, $i \leq m$, where μ is real and positive and $|\omega_i| = 1$; $|\lambda_i| < \mu$, $i > m$. As $t \to \infty$, (5.10) yields

$$\underset{\sim}{x}(t) \sim \mu^t \sum_{j=1}^{m} c_j \omega_j^t \underset{\sim}{V}_j. \tag{5.12}$$

Usually, we shall have $m = 1$, and $x_i(t)$ will decay or grow like μ^t.

If B does not have a complete set of eigenvectors, let the distinct eigenvalues $\lambda_1, \ldots, \lambda_k$ have multiplicities $r_1, \ldots, r_k$. We can always decompose $\underset{\sim}{x}(0)$ as

$$\underset{\sim}{x}(0) = \sum_{i=1}^{k} \underset{\sim}{U}_i, \tag{5.13}$$

where

$$(B - \lambda_i I)^{r_i} \underset{\sim}{U}_i = 0, \qquad i = 1, 2, \ldots k, \tag{5.14}$$

and I is the $n \times n$ identity matrix (see, e.g., Brauer and Nohel, 1969, p. 64). Substituting (5.13) into (5.8) and employing (5.14) gives

$$\underset{\sim}{x}(t) = \sum_{i=1}^{k} B^t \underset{\sim}{U}_i$$

$$= \sum_{i=1}^{k} [\lambda_i I + (B - \lambda_i I)]^t \underset{\sim}{U}_i$$

$$= \sum_{i=1}^{k} \sum_{j=0}^{t} \binom{t}{j} \lambda_i^{t-j} (B - \lambda_i I)^j \underset{\sim}{U}_i$$

$$= \sum_{i=1}^{k} \lambda_i^t \left[\sum_{j=0}^{r_i-1} \binom{t}{j} \lambda_i^{-j} (B - \lambda_i I)^j \right] \underset{\sim}{U}_i. \qquad (5.15)$$

The elements of the matrix in the brackets in (5.15) are polynomials of degree r_i-1. Thus, in general, polynomials may multiply the powers λ_i^t of degenerate eigenvalues. If $r_i = 1$ for all i, (5.15) reduces to (5.10). As $t \to \infty$, we may restrict the sum over i in (5.15) to the eigenvalue(s) of greatest absolute value.

Having developed the above machinery, we return to (5.6), with B given by (5.7). The eigenvalues are $\lambda_1 = w_1$, $\lambda_2 = \frac{1}{2}w_2$, $\lambda_3 = w_3$; the corresponding eigenvectors read

$$\underset{\sim}{V}_1 = \begin{pmatrix} 1 \\ 0 \\ 0 \end{pmatrix}, \quad \underset{\sim}{V}_2 = \begin{pmatrix} \lambda_2/[2(\lambda_2 - \lambda_1)] \\ 1 \\ \lambda_2/[2(\lambda_2 - \lambda_3)] \end{pmatrix}, \quad \underset{\sim}{V}_3 = \begin{pmatrix} 0 \\ 0 \\ 1 \end{pmatrix},$$

where $\underset{\sim}{V}_2$ does not exist if $\lambda_2 = \lambda_1$ or $\lambda_2 = \lambda_3$. If $\lambda_2 \neq \lambda_1, \lambda_3$, (5.10) applies, and (5.9) yields

$$c_1 = (1 \quad -\lambda_2/[2(\lambda_2 - \lambda_1)] \quad 0)\underset{\sim}{x}(0),$$

$$c_2 = (0 \quad 1 \quad 0)\underset{\sim}{x}(0),$$

$$c_3 = (0 \quad -\lambda_2/[2(\lambda_2 - \lambda_3)] \quad 1)\underset{\sim}{x}(0).$$

Assuming, without loss of generality, that $\lambda_1 \geq \lambda_3$, we have the following seven cases.

1. $\lambda_1 > \lambda_2 > \lambda_3$: From (5.4), (5.10), and the eigenvectors we infer at once that $y_1(t) \to 1$, i.e., AA is fixed, at the ultimate rate $(\lambda_2/\lambda_1)^t$.

2. $\lambda_1 > \lambda_3 > \lambda_2$: This is the same as Case 1, except that the asymptotic rate of convergence is $(\lambda_3/\lambda_1)^t$.

3. $\lambda_2 > \lambda_1 \geq \lambda_3$: The genotypic frequencies converge to $b\underset{\sim}{V}_2$, where the normalization constant

$$b = \frac{2(\lambda_2 - \lambda_1)(\lambda_2 - \lambda_3)}{4\lambda_2^2 - 3\lambda_2(\lambda_1 + \lambda_3) + 2\lambda_1\lambda_3}.$$

Thus, all three genotypes survive with frequencies $y_i(\infty) = bV_{2i}$. The ultimate rate of convergence is $(\lambda_1/\lambda_2)^t$.

4. $\lambda_1 = \lambda_3 > \lambda_2$: Aa is eliminated at the rate $(\lambda_2/\lambda_1)^t$, which simplifies to $(1/2)^t$ in the absence of selection. The ultimate ratio of the frequencies of AA and aa zygotes is c_1/c_3. With no selection, this reduces correctly to $p(0)/q(0)$.

Since the remaining situations have either $\lambda_2 = \lambda_1$ or $\lambda_2 = \lambda_3$ or both, we must use (5.15) or return to the original system (5.5). In view of the simplicity of (5.5), we shall do the latter. From (5.5b) we obtain

$$x_2(t) = x_2(0)\lambda_2^t, \tag{5.16}$$

whence (5.5c) gives

$$x_3' - \lambda_3 x_3 = \frac{1}{2}x_2(0)\,\lambda_2^{t+1}. \tag{5.17}$$

This is an inhomogeneous linear difference equation. It will be useful to examine linear equations of arbitrary order and dimension.

Suppose L is a linear operator and the vector $\underset{\sim}{z}(t)$ satisfies

$$L\underset{\sim}{z}(t) = \underset{\sim}{g}(t), \tag{5.18}$$

where $\underset{\sim}{g}(t)$ is known. Examples in one dimension in discrete and continuous time are

$$Lz(t) = z(t+3) - t^3 z(t+2) + 7z(t), \tag{5.19a}$$

$$Lz(t) = 2\frac{d^2z}{dt^2} + e^t\frac{dz}{dt} - z. \tag{5.19b}$$

In many dimensions, the left-hand side of (5.18) would be a vector with elements like (5.19). Assume we know a particular solution, $\underset{\sim}{z}_p(t)$, of (5.18) and the general solution, $\underset{\sim}{z}_h(t)$, of the homogeneous equation

$$L\underset{\sim}{z}_h(t) = 0. \tag{5.20}$$

If $\underset{\sim}{z}(t)$ is any solution (5.18), we have

$$L(\underset{\sim}{z} - \underset{\sim}{z}_p) = L\underset{\sim}{z} - L\underset{\sim}{z}_p = \underset{\sim}{g} - \underset{\sim}{g} = 0. \qquad (5.20)$$

Comparing (5.20) and (5.21), we conclude

$$\underset{\sim}{z}(t) = \underset{\sim}{z}_h(t) + \underset{\sim}{z}_p(t). \tag{5.22}$$

Returning to (5.17), we try the particular solution $\alpha\lambda_2^t$ to obtain

$$\alpha = \frac{\lambda_2 x_2(0)}{2(\lambda_2 - \lambda_3)}, \qquad \lambda_2 \neq \lambda_3.$$

For $\lambda_2 = \lambda_3$, motivated by (5.15), we substitute $\beta t\lambda_2^t$ and find

$$\beta = \frac{1}{2}x_2(0), \qquad \lambda_2 = \lambda_3.$$

The general solution of the homogeneous equation is obviously $\gamma\lambda_3^t$. Adding this to the particular solution and determining γ from the initial condition $x_3(0)$, we arrive at

$$x_3(t) = \begin{cases} \alpha\lambda_2^t + [x_3(0) - \alpha]\lambda_3^t, & \lambda_2 \neq \lambda_3, \\ \\ [\beta t + x_3(0)]\lambda_2^t, & \lambda_2 = \lambda_3. \end{cases} \tag{5.23}$$

From (5.5a) and (5.23), we deduce by symmetry

$$x_1(t) = \begin{cases} \bar{\alpha}\lambda_2^t + [x_1(0) - \bar{\alpha}]\lambda_1^t, & \lambda_2 \neq \lambda_1, \\ \\ [\beta t + x_1(0)]\lambda_2^t, & \lambda_2 = \lambda_1, \end{cases} \tag{5.24}$$

where

$$\bar{\alpha} = \frac{\lambda_2 x_2(0)}{2(\lambda_2 - \lambda_1)}.$$

The remaining results follow from (5.16), (5.23), and (5.24).

5. $\lambda_1 > \lambda_2 = \lambda_3$: As in Cases 1 and 2, *AA* is ultimately fixed, i.e., $y_1(\infty) = 1$, but the asymptotic rate of convergence is now $t(\lambda_2/\lambda_1)^t$. Heterozygotes, however, disappear faster, at the geometric rate $(\lambda_2/\lambda_1)^t$.

6. $\lambda_1 = \lambda_2 > \lambda_3$: Again $y_1(t) \to 1$, but now *Aa* and *aa* disappear at the slow algebraic rate $1/t$.

7. $\lambda_1 = \lambda_2 = \lambda_3$: Heterozygotes are lost at the rate $1/t$, and $y_1(\infty) = y_3(\infty) = 1/2$.

The approach presented here is closest to that of Karlin (1968), who reviews some of the literature.

5.2 *Assortative Mating with Multiple Alleles and Distinguishable Genotypes*

We suppose that all unordered genotypes have a different phenotype. A fraction r $(0 \le r \le 1)$ of each unordered genotype chooses a mate of its own genotype, while the remainder mates at random. Clearly, this model applies equally to a population with a proportion r of selfing and $1-r$ of random mating. We assume that all genotypes have the same fitness.

In our usual notation for ordered genotypic and mating frequencies,

$$P'_{ij} = \sum_{kl} X_{ik,lj}. \tag{5.25}$$

The contribution to $X_{ik,lj}$ from the panmictic subpopulation is $(1-r)P_{ik}P_{lj}$. In the assortatively mating subpopulation, $A_iA_i \times A_iA_i$ homozygote matings contribute rP_{ii}. The frequency of assortative, unordered $A_iA_k \times A_iA_k$ $(i \ne k)$ heterozygote matings is $2P_{ik}$. Inserting a factor of 1/2 to order each genotype, we obtain the contribution $r(1/2)^2(2P_{ik}) = \frac{1}{2}rP_{ik}$. Appropriate Kronecker deltas enable us to collect these frequencies in the single formula

$$X_{ik,lj} = (1-r)P_{ik}P_{lj} + r[\delta_{ik}\delta_{lj}\delta_{il}P_{ii} + (1-\delta_{ik})(\delta_{il}\delta_{kj} + \delta_{ij}\delta_{kl})(\tfrac{1}{2})P_{ik}]$$

$$= (1-r)P_{ik}P_{lj} + \frac{1}{2}r(\delta_{il}\delta_{kj} + \delta_{ij}\delta_{kl})P_{ik}. \tag{5.26}$$

It is easy to see that (5.26) satisfies (5.2), so we know that the gene frequencies, p_i, will be conserved. Substituting (5.26) into (5.25) yields the recursion relations

$$P'_{ij} = (1-r)p_ip_j + \frac{1}{2}r(P_{ij} + \delta_{ij}p_i), \tag{5.27}$$

which lead at once to $p'_i = p_i$. With $p_i = p_i(0)$, the solution of (5.27) is trivial. At equilibrium, the genotypic proportions are

$$\hat{P}_{ij} = \frac{2(1-r)p_ip_j + rp_i\delta_{ij}}{2-r}, \tag{5.28}$$

and the time dependence is given by

$$P_{ij}(t) = \hat{P}_{ij} + [P_{ij}(0) - \hat{P}_{ij}](\tfrac{1}{2}r)^t. \tag{5.29}$$

Using $Q_{ij}(t) = P_{ij}(t) - p_ip_j$ as indices of deviation from Hardy-Weinberg proportions, we find from (5.28) the excess of homozygotes,

$$\hat{Q}_{ii} = \frac{rp_i(1-p_i)}{2-r} > 0,$$

and the deficiency of heterozygotes,

$$\hat{Q}_{ij} = -\frac{rp_ip_j}{2-r} < 0, \qquad i \neq j.$$

From (5.28) and (5.29) we derive the heterozygosity

$$h(t) = \sum_{ij,\ i\neq j} P_{ij} = \hat{h} + [h(0) - \hat{h}](\tfrac{1}{2}r)^t,$$

where the equilibrium heterozygosity reads

$$\hat{h} = \left[\frac{2(1-r)}{2-r}\right] \sum_{ij,\ i\neq j} p_ip_j. \tag{5.30}$$

If the population is initially in Hardy-Weinberg proportions, then the sum in (5.30) is just $h(0)$, and (5.30) reduces to a result of Wright (1921). If mating is random, $r = 0$, (5.28) and (5.29) yield $P_{ij}(t) = \hat{P}_{ij} = p_ip_j$, $t \geq 1$, as expected. For complete assortative mating, $r = 1$, $\hat{P}_{ij} = p_i\delta_{ij}$, i.e., heterozygotes are eliminated. The rate of convergence is $(1/2)^t$. For two alleles, these results were derived in Section 5.1 (Case 4).

5.3 *Assortative Mating with Two Alleles and Complete Dominance*

A, with frequency p, is completely dominant to a, which has frequency q. A proportion α of dominant phenotypes and β of recessive ones mate assortatively, while the remainder mate at random. The assortative and random matings have fertilities 1 and $b = 1-s$, $s < 1$. The model with $\alpha = \beta$ and $s = 0$ was formulated and analyzed by O'Donald (1960). Scudo and Karlin (1969) generalized it to $\alpha \neq \beta$ and Crow and Kimura (1970, pp. 162-164) to $\alpha \neq \beta$ and $s \neq 0$. We put $P = P_{11}$, $Q = P_{12}$, $R = P_{22}$, so that the frequency of dominants is

$$D = P + 2Q = 1 - R,$$

and

$$p = P + Q, \qquad q = Q + R.$$

The proportions of the three subpopulations and the frequency of A within each will appear as follows.

1. Assortative recessive, with frequency βR, has $p_1 = 0$.
2. Assortative dominant, with proportion αD, has gene frequency

$$p_2 = (P + Q)/D = p/D.$$

3. Random mating, has frequencies

$$T = 1 - \beta R - \alpha D,$$

$$p_3 = (1 - \alpha)(P + Q)/T = (1 - \alpha)p/T.$$

The genotypic proportions in the next generation are sums of contributions from each of the subpopulations, which mate at random within themselves. Hence,

$$XP' = \alpha D p_2^2 + bTp_3^2, \tag{5.31a}$$

$$XQ' = \alpha D p_2 q_2 + bTp_3 q_3, \tag{5.31b}$$

$$XR' = \beta R + \alpha D q_2^2 + bTq_3^2. \tag{5.31c}$$

We introduced X in order to normalize the genotypic frequencies in the next generation. From (5.31) and the requirement $P' + 2Q' + R' = 1$ we derive

$$X = 1 - sT. \tag{5.32}$$

As anticipated, X is the mean fertility of the population. Of course,

$$q_2 = 1 - p_2 = Q/D,$$

$$q_3 = 1 - p_3 = [(1 - \beta)R + (1 - \alpha)Q] / T.$$

Substituting for p_i and q_i, we reduce (5.31) to

$$XP' = p^2 \left\{ \frac{\alpha}{D} + \frac{b(1 - \alpha)^2}{T} \right\}, \tag{5.33a}$$

$$XQ' = p \left\{ \frac{\alpha Q}{D} + \frac{b(1 - \alpha)[(1 - \beta)R + (1 - \alpha)Q]}{T} \right\}, \tag{5.33b}$$

$$XR' = \beta R + \frac{\alpha Q^2}{D} + \frac{b[(1 - \beta)R + (1 - \alpha)Q]^2}{T}. \tag{5.33c}$$

Calculating

$$Xp' = X(P' + Q') = p[1 - s(1 - \alpha)]$$

produces the gene frequency change

$$p = s(\alpha - \beta)pR / X. \tag{5.34}$$

It suffices to consider $s \geq 0$; the conclusions with $s < 0$ will clearly be the reverse of those with $s > 0$. We have the following situations.

1. $s > 0$:

(a) $\alpha > \beta$:

Evidently, $\Delta p \geq 0$. For $p > 0$, $\Delta p = 0$ if and only if $R = 0$. But (5.33c) informs us that $R = 0$ implies $R' = 0$ if and only if $Q = 0$, in which case $p = 1$. Therefore, $\Delta p > 0$ for $0 < p < 1$, so that $p(t) \to 1$ as $t \to \infty$. Obviously, this means $P(t) \to 1$. The reason A is ultimately fixed is that assortative matings are more fertile than random ones and dominants mate more assortatively than recessives. Hence, dominants are favored over recessives.

(b) $\alpha < \beta$:

Manifestly, by the argument just presented, $p(t) \to 0$.

(c) $\alpha = \beta$:

In view of the biological argument above, it is hardly surprising that with equal assortative mating p is constant. We focus our attention on (5.33a); noting that $D = 2p - P$, $T = 1 - \alpha$, and $X = 1 - s(1 - \alpha)$, we simplify it to the linear fractional transformation

$$P' = \frac{p^2}{1 - s(1 - \alpha)} \left[\frac{\alpha}{2p - P} + (1 - s)(1 - \alpha) \right] \equiv f(P). \qquad (5.35)$$

Since $Q = p - P \geq 0$ implies $P \leq p$ and $R = q - Q = 1 - 2p + P \geq 0$ implies $P \geq 2p - 1$, (5.35) is a mapping of the interval $[P^*, p]$ into itself, where $P^* = \max(0, 2p - 1)$. It is easy to verify for $0 < p < 1$ that $f(P^*) > P^*$, and $f(p) < p$ unless $\alpha = 1$. From the theory of the linear fractional transformation developed in Section 2.2, it follows immediately that for $0 \leq \alpha < 1$ $P(t)$ converges globally to the unique root in (P^*, p) of the quadratic $f(P) = P$. (In fact, we have Case 2(a) of Section 2.2.)

If $\alpha = 1$, (5.35) reduces to

$$P' = p^2 / (2p - P). \qquad (5.36)$$

With the substitution $x = 1/(p - P)$, we easily derive

$$P(t) = p - \frac{p[p - P(0)]}{p + [p - P(0)]t},$$

showing that $P(t) \to p$, $Q(t) \to 0$, $R(t) \to q$ at the algebraic rate $1/t$. Heterozygotes are eliminated because assortative mating is complete.

2. $s = 0$:

With no fertility differences, (5.34) tells us that p is

constant, and (5.32) yields $X = 1$. Therefore, (5.33a) becomes

$$P' = p^2\left[\frac{\alpha}{2p-P} + \frac{(1-\alpha)^2}{1-\beta+(\beta-\alpha)(2p-P)}\right] \equiv g(P). \tag{5.37}$$

Again, $g(P)$ is a mapping of $[P^*,p]$ into itself. For $0 < p < 1$, we have $g(P^*) > P^*$, and $g(p) < p$ unless $\alpha = 1$ or $\beta = 1$. If $\alpha = 1$ or $\beta = 1$, (5.37) reduces to (5.36). Therefore, we assume $\alpha,\beta < 1$, and exclude also the trivial case of pure panmixia, $\alpha = \beta = 0$. From (5.37) we compute

$$\frac{dg}{dP} = p^2\left[\frac{\alpha}{(2p-P)^2} + \frac{(\beta-\alpha)(1-\alpha)^2}{[1-\beta+(\beta-\alpha)(2p-P)]^2}\right], \tag{5.38}$$

$$\frac{d^2g}{dP^2} = 2p^2\left[\frac{\alpha}{(2p-P)^3} + \frac{(\beta-\alpha)^2(1-\alpha)^2}{[1-\beta+(\beta-\alpha)(2p-P)]^3}\right] > 0. \tag{5.39}$$

The convexity displayed by (5.39) implies

$$\frac{dg}{dP}(P) \geq \frac{dg}{dP}(P^*),$$

with equality only at $P = P^*$.

We shall show that $g(P)$ is increasing by proving that

$$\frac{dg}{dP}(P^*) \geq 0.$$

(a) $p \leq 1/2$:

Here $P^* = 0$, and (5.38) yields

$$\frac{dg}{dP}(0) = \frac{\alpha}{4} + \frac{p^2(\beta-\alpha)(1-\alpha)^2}{[1-\beta+2p(\beta-\alpha)]^2}. \tag{5.40}$$

If $\beta \geq \alpha$, obviously

$$\frac{dg}{dP}(0) > 0.$$

If $\beta < \alpha$, since $p \leq 1/2$, from (5.40) we obtain

$$\frac{dg}{dP}(0) \geq \frac{\alpha}{4} + \frac{(1/2)^2(\beta-\alpha)(1-\alpha)^2}{[1-\beta+2(1/2)(\beta-\alpha)]^2} = \frac{\beta}{4} \geq 0.$$

(b) $p > 1/2$:

Now $P^* = 2p-1$, and (5.38) gives

$$\frac{dg}{dP}(2p-1) = \beta p^2 \geq 0.$$

Collecting our results enables us to sketch $g(P)$ in Figure 5.1,

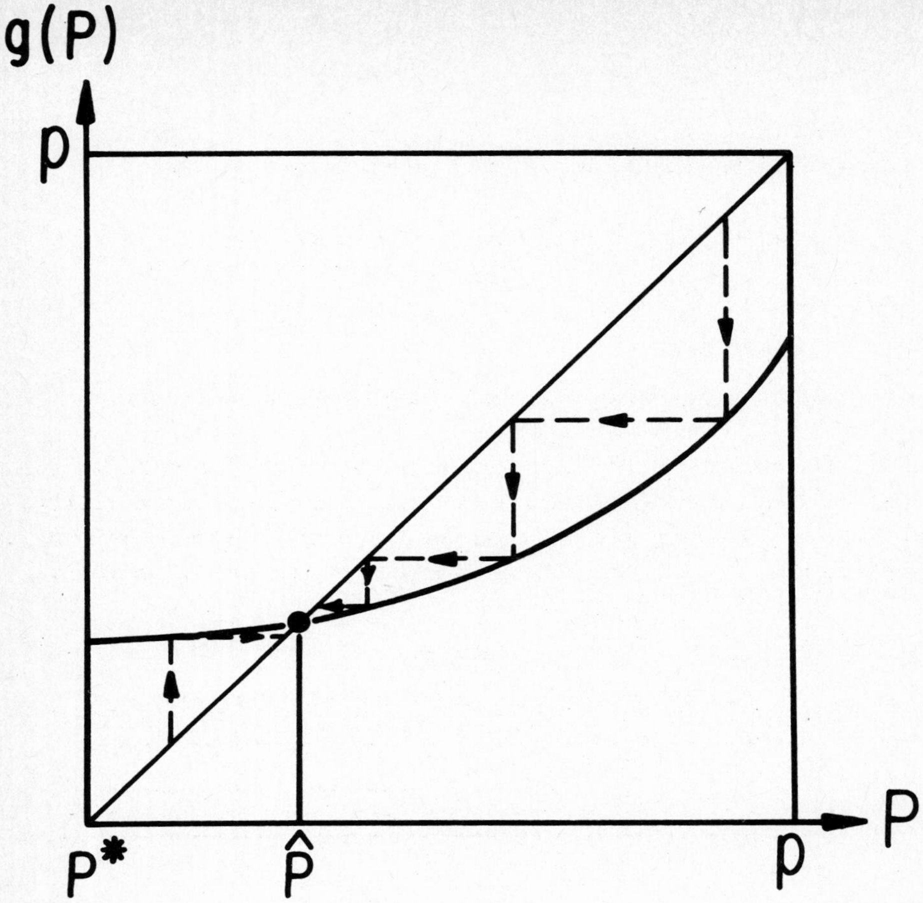

Fig. 5.1. The iteration of (5.37).

which shows immediately that $P(t)$ converges globally, without oscillation, to $\hat{P}$, the unique root in (P^*, p) of the cubic $g(P) = P$. Since mating is assortative, we expect an excess of homozygotes relative to Hardy-Weinberg proportions. We leave it for Problem 5.3 to prove that $\hat{P} > p^2$. If $\alpha = \beta$, (5.37) reduces to the special case of (5.35) with $s = 0$.

Finally, we shall show that the departure from Hardy-Weinberg proportions is approximately linear in α and β for $\alpha, \beta << 1$. Expanding (5.37) leads to

$$\frac{g(P)}{p^2} = 1 + \frac{(1-2p+P)[\alpha(1-2p+P) + \beta(2p-P)]}{2p-P} + O(\alpha^2,\alpha\beta,\beta^2). \qquad (5.41)$$

Therefore,

$$\hat{P} = g(\hat{P}) = p^2 + O(\alpha,\beta). \qquad (5.42)$$

Substituting (5.42) into (5.41) gives

$$\frac{\hat{P}}{p^2} = 1 + \frac{q^2[\alpha q^2 + \beta p(1+q)]}{p(1+q)} + O(\alpha^2,\alpha\beta,\beta^2). \qquad (5.43)$$

For sufficiently small α and β, (5.43) proves that $\hat{P} > p^2$.

5.4. *Random Mating with Differential Fertility*

A is again dominant to a; p, q, P, Q, R, D are as in the previous section. Mating is random, but dominant × recessive matings have fertility $b = 1 - s$, $0 < s < 1$, instead of unity. The difference equations are most easily written by collecting progeny in Table 5.1.

TABLE 5.1

Random mating with differential fertility

Mating	Contribution	Progeny		
		AA	Aa	aa
$AA \times AA$	P^2	1	0	0
$AA \times Aa$	$2P(2Q)$	1/2	1/2	0
$AA \times aa$	$2PRb$	0	1	0
$Aa \times Aa$	$(2Q)^2$	1/4	1/2	1/4
$Aa \times aa$	$2(2Q)Rb$	0	1/2	1/2
$aa \times aa$	R^2	0	0	1

The total contribution is just the mean fecundity,

$$X = 1 - 2sRD. \qquad (5.44)$$

The recursion relations

$$XP' = P^2 + 2PQ + Q^2,$$

$$XQ' = PQ + PRb + Q^2 + QRb,$$

$$XR' = Q^2 + 2QRb + R^2,$$

directly simplify to

$$XP' = p^2, \tag{5.45a}$$

$$XQ' = p(q - sR), \tag{5.45b}$$

$$XR' = q^2 - 2sQR. \tag{5.45c}$$

From (5.45a) and (5.45b) we deduce

$$Xp' = p(1 - sR). \tag{5.46}$$

We employ p and R as our two independent variables, and begin by locating the equilibria. If $p' = p$, (5.46) yields $p = 0$, which implies $R = 1$; or

$$X = 1 - sR. \tag{5.47}$$

Equating (5.44) and (5.47) immediately gives $R = 0$ or $\hat{D} = \frac{1}{2}$. At equilibrium, setting $R = 0$ in (5.45c) informs us that $q = 0$. So, this is just the other trivial equilibrium. To find the gene frequency at the polymorphic equilibrium, we substitute $\hat{D} = \frac{1}{2}$ into (5.45a):

$$(1 - \tfrac{1}{2}s)(2p - \tfrac{1}{2}) = p^2.$$

This leads to

$$\hat{p} = \tfrac{1}{2}[2 - s - \sqrt{(2 - s)(1 - s)}\,]. \tag{5.48a}$$

The root with a positive sign in front of the radical is trivially seen to exceed 1/2, which is impossible because $\hat{p} \leq \hat{D} = 1/2$. Note that

$$\hat{q} = \tfrac{1}{2}[s + \sqrt{(2 - s)(1 - s)}\,]. \tag{5.48b}$$

We now turn to the evolution of the population. Since

$$P = p - Q = 2p - 1 + R \geq 0 \tag{5.49a}$$

and

$$Q = q - R = 1 - p - R \geq 0, \tag{5.49b}$$

the system is confined to the interior of the triangle with the heavily drawn boundaries in Figure 5.2. Dominant and recessive individuals have respective probabilities R and D of mating disassortatively. Hence, an individual with the more (less) frequent phenotype has a lower (higher) probability of mating disassortatively. Since disassortative matings are less fertile, the more frequent phenotype is favored. Therefore, we expect the two trivial equilibria to be stable and the polymorphic one to be unstable. Establishing the convergence

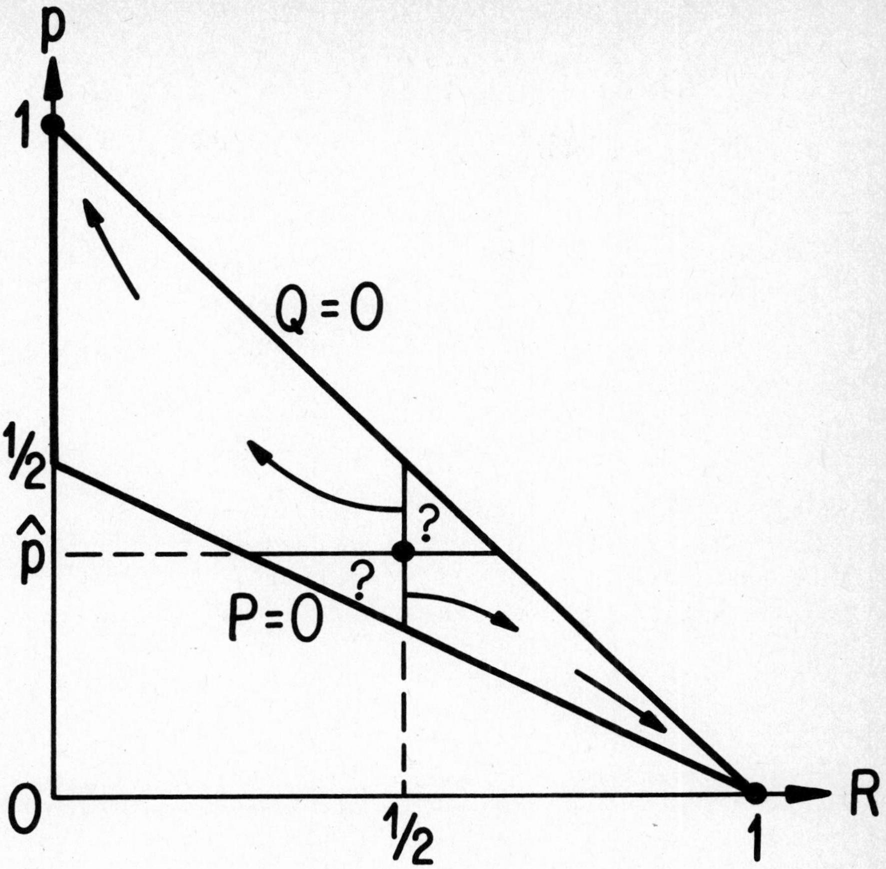

Fig. 5.2. The iteration of (5.45).

pattern displayed in Figure 5.2 proves considerably more than this. If the population starts in either of the two regions marked with a question mark, the analysis below does not determine its dynamics. The simplest possibility is the existence of a line, passing through the unstable equilibrium and the two *terrae incognitae*, separating the regions of attraction of the two stable equilibria. This need not apply, however; e.g., a population starting in *terra incognita* may not converge to an equilibrium at all.

We wish to prove that if $p(0) > \hat{p}$ and $R(0) \leq 1/2$, then $p(t) \to 1$ as $t \to \infty$. From (5.44) and (5.46) we have

$$p' = \frac{p(1 - sR)}{1 - 2sR(1 - R)}.$$

This shows that $p' \gtrless p$ for $R \lessgtr 1/2$. Hence, it suffices to demonstrate that $p > \hat{p}$ and $R \leq 1/2$ imply $R' < 1/2$, for then the population cannot enter the upper triangle with the question mark. We rewrite (5.45c) in the form

$$f(q,R) \equiv X(R' - \tfrac{1}{2}) = q^2 - \tfrac{1}{2} + sR(1 + R - 2q). \tag{5.50}$$

Now,

$$\frac{\partial f}{\partial R} = s(1 + R - 2q) + sR = s(P + R) > 0$$

from (5.49a). Therefore, $f(q,R)$ is monotone increasing in R for fixed q. In the region under consideration, $R \leq \min(q,1/2)$.

If $q \leq 1/2$,

$$f(q,R) \leq f(q,q) = (1 - s)q^2 + sq - \tfrac{1}{2} < 0$$

for $0 \leq q \leq 1/2$. If $q > 1/2$,

$$f(q,R) \leq f(q,1/2) = (q - \hat{q})(q + \hat{q} - s) \tag{5.51}$$

from (5.48b). But

$$q + \hat{q} - s > \tfrac{1}{2} + \hat{q} - s > 0,$$

the positivity following from the trite calculation ruling out a positive sign in front of the radical in (5.48a). Consequently, for $\frac{1}{2} < q < \hat{q}$ we have established that $f(q,R) < 0$. Together with the same result for $q \leq 1/2$, this proves that, as required, $R' < 1/2$ in the region of interest.

To study the dynamics in the quadrilateral on the lower right in Figure 5.2, we suppose $p(0) < \hat{p}$ and $R(0) \geq 1/2$. From the argument for the upper pentagon, we only need to show that $p < \hat{p}$ and $R \geq 1/2$ imply $R' > 1/2$. For $R \geq 1/2$, we have (5.51) with the inequality reversed. Since $q > \hat{q}$, (5.51) becomes

$$f(q,R) > (q - \hat{q})(2\hat{q} - s) > 0,$$

the positivity being ensured by (5.48b). Therefore, $R' > 1/2$, and the proof is complete.

Since with $R(0) = 1/2$, $p(t)$ tends to 0 or 1 according as $p(0) < \hat{p}$ or $p(0) > \hat{p}$, it is clear that $R = 1/2$ does not separate the domains of attractions of the two stable equilibria. Therefore, one cannot take the intuitive argument with which we commenced our dynamical investigation completely literally.

We owe the model discussed in this section to Scudo and Karlin (1969); our interpretation is that of Crow and Kimura (1970, pp. 164-166). Down to (5.50) our analysis followed Karlin (1968).

Let us end this section by calculating the ultimate rates of convergence to the pure states.

1. $p(\infty) = 0$:

Since $D(\infty) = 0$, we use D instead of R. Recalling (5.44), to first order in p and D, (5.46) becomes

$$p' \sim (1-s)p,$$

whence

$$p(t) \sim k(1-s)^t \tag{5.52}$$

as $t \to \infty$, for some constant k. To first order, we deduce from (5.45c)

$$D' \sim 2(1-s)p.$$

Substituting (5.52) yields

$$D(t) \sim 2(1-s)p(t-1) \sim 2p(t) \tag{5.53}$$

as $t \to \infty$. Thus, the dominant allele, A, is lost at a geometric rate.

2. $p(\infty) = 1$:

The convenient variables now are q and R. From (5.46) and (5.45c) we obtain

$$q' = q - sR + O(qR, R^2), \tag{5.54a}$$

$$R' = q^2 - 2sR(q-R) + O(q^2R, qR^2, R^3). \tag{5.54b}$$

On the basis of (5.54b) we expect $R' \sim q^2$ as $t \to \infty$. Hence, (5.54a) yields

$$\frac{q''}{q'} \sim 1 - \frac{sq^2}{q'} \to 1$$

as $t \to \infty$, so that $R \sim q^2$, and q satisfies

$$q' \sim q(1-sq).$$

With the method applied to (4.39) we obtain (see Problem 5.4)

$$q(t) = \frac{1}{st} - \frac{\ln t}{st^2} + O(t^{-2}) \tag{5.55a}$$

as $t \to \infty$. Therefore, $R \sim q^2$ yields

$$R(t) = \frac{1}{s^2t^2} - \frac{2\ln t}{s^2t^3} + O(t^{-3}). \tag{5.55b}$$

It is easy to check that (5.55) satisfies (5.54) with the errors indicated. As in Section 4.2, the recessive allele is lost at an algebraic rate.

5.5 *Self-Incompatibility Alleles*

Self-incompatibility systems in plants usually increase genetic variability by preventing self-fertilization. Many flowering plants possess a self-incompatibility locus such that pollen will not function on a style carrying the allele in the pollen grain. This mechanism evidently favors rare alleles. The extensive literature on the prediction of the expected number of alleles present in terms of the population size and the mutation rate is briefly reviewed in Nagylaki (1975a). In this section, we shall concentrate on one component of this interesting and difficult problem: the deterministic behavior of the system (Nagylaki, 1975a).

We denote the frequencies of the n self-incompatibility alleles, S_i, by p_i, and those of the ordered genotypes, S_iS_j, by P_{ij}. Since S_i and S_j pollen will not fertilize an S_iS_j plant, there are no homozygotes: $P_{ii} = 0$. Clearly, $n \geq 3$. Let us suppose that the proportion of S_iS_j genotypes among fertilized plants is P_{ij}. This means that no failures of fertilization are to be ascribed to the different allelic frequencies in pollen, a reasonable assumption if there is excess pollen. The frequency of S_k, $k \neq i,j$, among pollen which can fertilize S_iS_j is $p_k/(1-p_i-p_j)$. Positing random pollination, this argument yields Fisher's (1958) equation,

$$2P'_{ij} = \sum_{k\neq i,j} \frac{P_{ik}p_j}{1-p_i-p_k} + \sum_{k\neq i,j} \frac{P_{kj}p_i}{1-p_k-p_j}, \tag{5.56}$$

for $i \neq j$. Since there are no homozygotes, obviously, $p_i \leq 1/2$ for all i. If $p_i = p_j = 1/2$ for any particular i,j, all genotypes must be S_iS_j, leading to extinction of the population. Thus, the denominators in (5.56) are positive. From (5.56) we deduce the gene frequency equation

$$2p'_i = p_i + \sum_{j\neq i} \sum_{k\neq i,j} \frac{P_{kj}p_i}{1-p_k-p_j}. \tag{5.57}$$

Another summation verifies that the normalization is correct.

It is easy to check that the symmetry point

$$\hat{P}_{ij} = 1/[n(n-1)], \qquad \hat{p}_i = 1/n \tag{5.58}$$

is an equilibrium of (5.56). The results presented below support the conjecture that (5.58) is the only (completely polymorphic) equilibrium of (5.56), and (5.58) is globally asymptotically stable. Neither a proof nor a counterexample has been found.

For three alleles, and only for three alleles, (5.56) is linear, and a trite calculation left for Problem 5.5 gives

$$P_{ij}(t) = \frac{1}{6} + [P_{ij}(0) - \frac{1}{6}](-\frac{1}{2})^t. \tag{5.59}$$

The allelic frequencies may be computed from

$$p_k(t) = \frac{1}{2} - P_{ij}(t) \qquad k \neq i,j. \tag{5.60}$$

Equations (5.59) and (5.60) show that the genotypic and allelic frequencies converge globally to the symmetry point, oscillating at the rate $(-1/2)^t$.

We shall now demonstrate that the symmetric equilibrium (5.58) is asymptotically stable. Consider the transformation

$$\underset{\sim}{x}' = \underset{\sim}{f}(\underset{\sim}{x}) \tag{5.61}$$

of a set of variables x_i. Suppose $\hat{\underset{\sim}{x}}$ is an equilibrium:

$$\hat{\underset{\sim}{x}} = \underset{\sim}{f}(\hat{\underset{\sim}{x}}). \tag{5.62}$$

Near $\underset{\sim}{x} = \hat{\underset{\sim}{x}}$ (5.61) has the Taylor expansion

$$x_i' = f_i(\hat{\underset{\sim}{x}}) + \sum_i \frac{\partial f_i}{\partial x_j}(\hat{\underset{\sim}{x}})(x_j - \hat{x}_j) + O(|\underset{\sim}{x} - \hat{\underset{\sim}{x}}|^2). \tag{5.63}$$

Defining $\underset{\sim}{\varepsilon} = \underset{\sim}{x} - \hat{\underset{\sim}{x}}$ and

$$b_{ij} = \frac{\partial f_i}{\partial x_j}(\hat{\underset{\sim}{x}}), \tag{5.64a}$$

and substituting (5.62) into (5.63) produces the linearized equation

$$\underset{\sim}{\varepsilon}' = B\underset{\sim}{\varepsilon}. \tag{5.64b}$$

Glancing at (5.15), we expect $\hat{\underset{\sim}{x}}$ to be asymptotically stable if the absolute value of all the eigenvalues of B is less than unity. This assertion can be justified rigorously (Bellman, 1949).

To apply the above procedure to our problem, put

$$P_{ij} = \hat{P}_{ij} + \varepsilon_{ij}, \qquad p_i = \hat{p}_i + \eta_i, \tag{5.65}$$

with

$$\eta_i = \sum_{j \neq i} \varepsilon_{ij}, \qquad \sum_i \eta_i = 0. \tag{5.66}$$

Substituting (5.65) and (5.66) into (5.56), we find the linearized equation

$$\varepsilon'_{ij} = -\mu_n \varepsilon_{ij} + \left(\frac{\eta_i + \eta_j}{n-2}\right)\left(1 - \frac{1}{(n-2)(n-1)}\right), \tag{5.67}$$

where

$$\mu_n = 1/(n-2). \tag{5.68a}$$

From (5.66) and (5.67) we derive $\eta'_i = \lambda_n \eta_i$, with

$$\lambda_n = 1 - \frac{n}{(n-2)(n-1)}. \tag{5.68b}$$

Therefore,

$$\eta_i(t) = \eta_i(0)\lambda_n^t. \tag{5.69}$$

Inserting (5.69) into (5.67) yields a difference equation of the same form as (5.17); its solution reads

$$\varepsilon_{ij}(t) = \mu_n[\eta_i(0) + \eta_j(0)]\lambda_n^t + \{\varepsilon_{ij}(0) - \mu_n[\eta_i(0) + \eta_j(0)]\}(-\mu_n)^t. \tag{5.70}$$

It must be kept in mind that (5.70) is a good approximation only if $|\varepsilon_{ij}(0)| << \hat{P}_{ij}$. The case of three alleles is an exception: since (5.56) is linear, therefore (5.70) is exact, and is easily seen to reduce to (5.59). Hence, (5.68) and (5.70) show that $\varepsilon_{ij}(t) \to 0$ as $t \to \infty$.

The nature of the convergence to equilibrium is rather interesting. For $n = 3$, both the genotypic and the gene frequencies oscillate at the rate $(-1/2)^t$. If there are four alleles, $\mu_4 = 1/2 > 1/3 = \lambda_4$, so (5.69) and (5.70) inform us that ultimately the genotypic frequencies oscillate, but the gene frequencies do not. With $n \geq 5$, $\lambda_n > \mu_n$, giving nonoscillatory asymptotic convergence at the rate λ_n^t. In the biologically important case of many alleles, $n >> 1$, $\lambda_n \approx 1 - n^{-1}$, and the rate of convergence is very close to $e^{-t/n}$, which has a characteristic time equal to the number of alleles.

We shall briefly consider the fate of rare alleles. From (5.57) it can be proved that (Nagylaki, 1975a)

$$\Delta p_i > p_i[-p_i(2 - p_i) + (1 - p_i)^3(n - 2 + p_i)^{-1}].$$

The bracket is positive at $p_i = 0$, and its smallest positive root is

$$q = 1 + a - (1 + a^2)^{1/2},$$

where $a = [2(n-1)]^{-1}$. Consequently, $\Delta p_i > 0$ for $0 < p_i < q$, showing that rare alleles are established. Note that $q > 1/(2n)$, but $q \approx 1/(2n)$ if $n >> 1$. The condition $p_i < q$ for $\Delta p_i > 0$ is sufficient, but not necessary. Numerical counterexamples for four alleles dispose of the naive conjecture that $\Delta p_i > 0$ for $p_i < 1/n$ (Nagylaki, 1975a).

5.6 *Pollen and Zygote Elimination*

As our final example, we shall examine some disassortative mating schemes proposed for plants by Finney (1952). He assumed that only crosses of unlike phenotypes could produce offspring, and formulated two models for random pollination which differ in the specification of the phenotypic mating frequencies.

Let r_i and R_{ij} ($R_{ii} = 0$) represent the frequencies of the phenotype π_i and the ordered union $\pi_i \times \pi_j$. The proportion of pollen from π_i plants among pollen which can fertilize π_j is $r_i/(1-r_j)$. If there are no failures of fertilization due to the phenotypic origin of the pollen, the frequency of π_j phenotypes among fertilized plants is r_j. This should be a good approximation with excess pollen. Thus, the *pollen elimination* model is based on the ordered mating frequencies

$$R_{ij} = \tfrac{1}{2}r_j[r_i/(1-r_j)] + \tfrac{1}{2}r_i[r_j/(1-r_i)] = \tfrac{1}{2}r_i r_j[(1-r_i)^{-1} + (1-r_j)^{-1}] \tag{5.71}$$

for $i \neq j$. It is easy to verify that (5.71) is correctly normalized. Note that the reasoning used to deduce (5.56) for incompatibility determined by the allele in the pollen and the plant genotype corresponds to pollen elimination.

If pollen is relatively scarce, we suppose that seeds from incompatible crosses do not germinate. Then the ordered mating frequencies read

$$R_{ij} = r_i r_j / X, \qquad i \neq j, \tag{5.72a}$$

where

$$X = \sum_{ij} R_{ij} = 1 - \sum_i r_i^2. \tag{5.72b}$$

Following Finney (1952), we shall call (5.72) *zygote elimination*.

With only two phenotypes, since there is but one fertile mating, pollen and zygote elimination are equivalent.

We proceed to analyze (5.71) and (5.72) for a single diallelic locus. It will be convenient to use P, H, R for the frequencies of AA, Aa, aa; the heterozygote frequency, $H = 2P_{12}$, is unordered here.

1. Complete Dominance

Suppose A is dominant to a. As explained above, pollen and zygote elimination are the same. The permitted crosses, $AA \times aa$ and $Aa \times aa$, have frequencies proportional to P and H. Therefore, our recursion relations are

$$TP' = 0, \tag{5.73a}$$

$$TH' = P + \tfrac{1}{2}H, \tag{5.73b}$$

$$TR' = \tfrac{1}{2}H \tag{5.73c}$$

where normalization requires

$$T = P + H. \tag{5.74}$$

From (5.73a), we have $P(t) = 0$, $t \geq 1$. Of course, the elimination of dominant homozygotes in one generation is quite obvious. Hence, (5.74) gives $T' = H'$. Substituting this into (5.73b) informs us that $H'H'' = \frac{1}{2}H'$. But (5.73b) also shows that $H' = 0$ only if $R = 1$, leading to extinction of the population in one generation, so $H'' = 1/2$. Thus, the equilibrium $R = H = 1/2$ is attained in two generations (Finney, 1952).

2. Identical Homozygotes

Again, there is only one scheme. Now the fertile unions, $AA \times Aa$ and $aa \times Aa$, have frequencies proportional to P and R, yielding the difference equations

$$TP' = \tfrac{1}{2}P, \tag{5.75a}$$

$$TH' = \tfrac{1}{2}P + \tfrac{1}{2}R, \tag{5.75b}$$

$$TR' = \tfrac{1}{2}R, \tag{5.75c}$$

where

$$T = P + R. \tag{5.76}$$

Adding (5.75a) and (5.75c) and excluding $H = 1$ to prevent extinction in a single generation, we obtain $P' + R' = 1/2$, i.e., $H(t) = 1/2$, $t \geq 1$. Dividing (5.75a) by (5.75c) shows that $P(t)/R(t)$ is constant, whence we derive for $t \geq 1$

$$P(t) = \tfrac{1}{2}P(0)/[P(0)+R(0)], \qquad R(t) = \tfrac{1}{2}R(0)/[P(0)+R(0)]. \tag{5.77}$$

This equilibrium is reached in just one generation, but depends on the initial conditions (Finney, 1952).

3. Pollen Elimination with Distinct Genotypes

If all three genotypes have distinguishable phenotypes, pollen and zygote elimination lead to different models with quite dissimilar dynamical behavior. The permitted matings are $AA \times Aa$, $AA \times aa$, and $Aa \times aa$. The equations for pollen elimination follow directly from (5.71) and Mendelian segregation:

$$P' = \tfrac{1}{2}PH[(1-P)^{-1}+(1-H)^{-1}],$$

$$H' = \tfrac{1}{2}PH[(1-P)^{-1}+(1-H)^{-1}]+PR[(1-P)^{-1}+(1-R)^{-1}]+\tfrac{1}{2}HR[(1-H)^{-1}+(1-R)^{-1}],$$

$$R' = \tfrac{1}{2}HR[(1-H)^{-1}+(1-R)^{-1}].$$

This system reduces to

$$P' = \frac{PH(1+R)}{2(1-P)(1-H)}, \qquad (5.78a)$$

$$H' = \frac{1}{2} + \frac{PR(1+H)}{2(1-P)(1-R)}, \qquad (5.78b)$$

$$R' = \frac{HR(1+P)}{2(1-H)(1-R)}. \qquad (5.78c)$$

We cannot allow any genotypic frequency to be unity because the population would be extinguished in the next generation. We conclude at once from (5.78b) that the frequency of heterozygotes will be at least 1/2 after one generation. Hence, for $R \neq 0$ the ratio of (5.78a) to (5.78c) gives

$$\frac{P'}{R'} = \frac{P(1-R^2)}{R(1-P^2)}. \qquad (5.79)$$

Let us locate the equilibria of (5.78). If $R \neq 0$, (5.79) implies $\hat{P} = \hat{R}$ or $P = 0$. Assuming $P \neq 0$ and substituting $\hat{P} = \hat{R} = \frac{1}{2}(1-\hat{H})$ into (5.78b) at equilibrium, we deduce

$$\hat{H}^2 + 3\hat{H} - 2 = 0,$$

whence

$$\hat{H} = (\sqrt{17} - 3)/2 = 0.562. \qquad (5.80)$$

From (5.78a) and (5.78b) we infer that $P = 0$, $H = 1/2$ (from which R = 1/2) is an equilibrium. Since (5.78) is symmetric under the interchange $P \leftrightarrow R$, it is not surprising that with $R = 0$ we find the

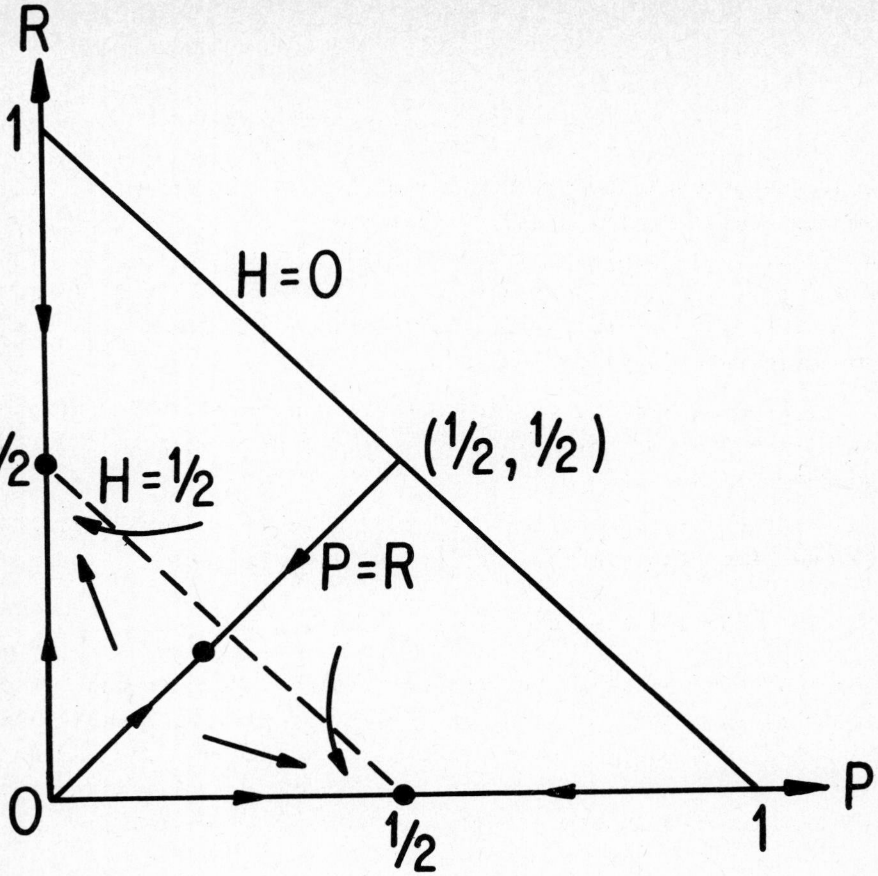

Fig. 5.3. The iteration of (5.78).

corresponding equilibrium, $H = 1/2$, $P = 1/2$. The three equilibria are shown in Figure 5.3.

We continue with the demonstration of the convergence pattern displayed in Figure 5.3. If $P = 0$, then $P' = 0$ and $H' = 1/2$; if $R = 0$, then $R' = 0$ and $H' = 1/2$. Henceforth, we may suppose $P(0) > 0$ and $R(0) > 0$. We may also assume $H(0) > 0$, for otherwise the population will be extinguished in the second generation. Therefore, we may analyze (5.78) for $P,H,R > 0$.

If $P < R$, we infer from (5.79) that $P'/R' < P/R$. Suppose $P(0) < R(0)$. Since the monotone decreasing sequence $\{P(t)/R(t)\}$ is bounded below by 0, it must have a limit, say α, $0 \le \alpha < 1$. If $\alpha > 0$, (5.79) yields

$$1 = \lim_{t\to\infty} \frac{1-[R(t)]^2}{1-[P(t)]^2} = \lim_{t\to\infty} \frac{1-[R(t)]^2}{1-\alpha^2[R(t)]^2}. \tag{5.81}$$

Hence, the sequence $\{R(t)\}$ has a limit, $R(\infty)$. In view of the equilibrium structure, $R(\infty) = 0$ would imply $P(\infty) = 1/2 > R(\infty)$. Thus, $R(\infty) > 0$, and (5.81) is impossible. Therefore, $\alpha = 0$ and $P(\infty) = 0$. Since $H(t) \ge 1/2$ for $t \ge 1$, $R(t)$ cannot tend to 1. Then (5.78b) informs us that $H(t) \to 1/2$. We conclude from (5.79) that P/R converges to 0 at the asymptotic rate $(3/4)^t$.

If $P(0) > R(0)$, the same argument applied to R/P demonstrates that $P(\infty) = H(\infty) = 1/2$, and shows that R/P converges to 0 at the ultimate rate $(3/4)^t$.

If $P = R$, according to (5.79), $P' = R'$. Therefore, if $P(0) = R(0)$, we have $P(t) = R(t)$, and (5.78b) simplifies to

$$H' = \frac{1}{2} + \frac{(1-H)^2}{2(1+H)} \equiv f(H), \qquad 0 \le H \le 1. \tag{5.82}$$

We exclude $H(0) = 0,1$ to prevent extinction. Observe that there is no H in $(0,1)$ such that $f(H) = 0$ or $f(H) = 1$. We note that $f(0) = 1$ and $f(1) = 1/2$, and easily verify that $f(H)$ is monotone decreasing for $0 \le H < 1$ and convex for $0 \le H \le 1$. Therefore, the mapping (5.82) has the form depicted in Figure 5.4, which has the crucial feature

$$H < f(H) < 2\hat{H} - H, \qquad 0 < H < \hat{H}, \tag{5.83a}$$

$$2\hat{H} - H < f(H) < H, \qquad \hat{H} < H < 1. \tag{5.83b}$$

Subtracting $\hat{H}$ from (5.83), we obtain

$$|f(H) - \hat{H}| < |H - \hat{H}|, \qquad 0 < H < 1.$$

It follows from Problem 4.5 that $H(t) \to \hat{H}$ globally. The convergence is clearly oscillatory. From (5.64), (5.80), and (5.82) we deduce the ultimate rate of convergence λ^t, where

$$\lambda = \frac{df}{dH}(\hat{H}) = -\frac{1}{\sqrt{17}-1} = -0.320.$$

Of course, in practice small perturbations would drive the population into the region of attraction of one of the two stable equilibria.

The equilibrium analysis of this model is due to Finney (1952). Our proof of convergence to the stable equilibria follows Karlin and Feldman (1968).

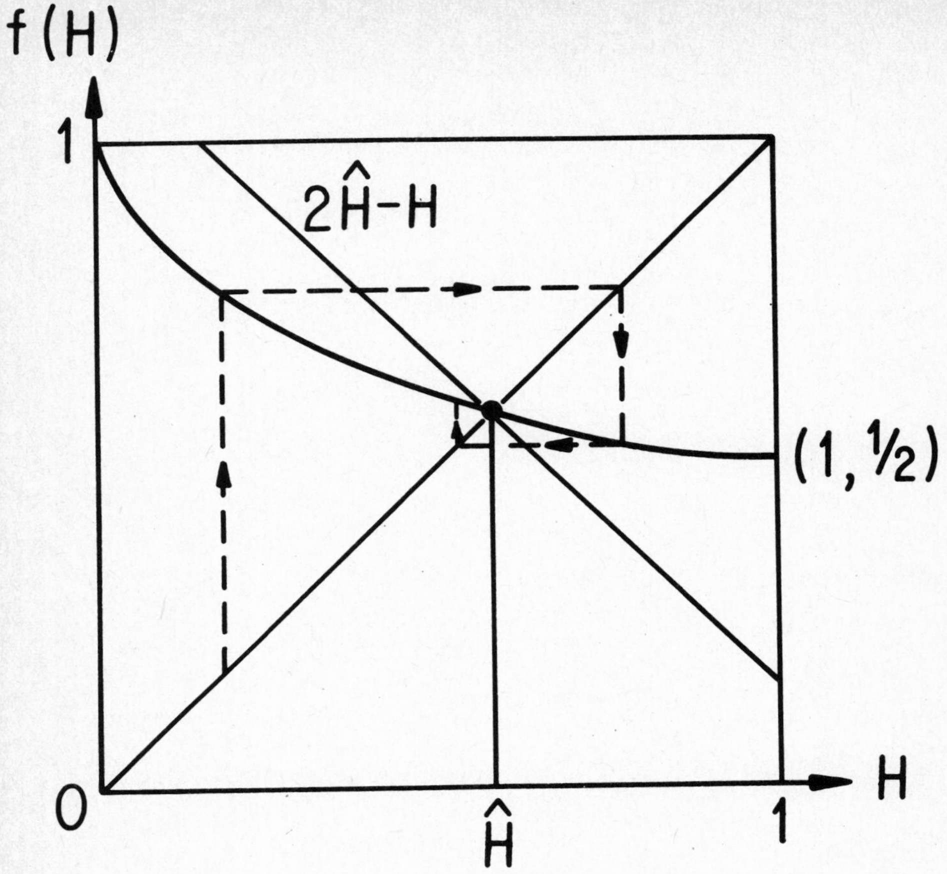

Fig. 5.4. The iteration of (5.82).

4. Zygote Elimination with Distinct Genotypes

The fertile crosses are the same as in the pollen elimination model above, but now we write our equations by specializing (5.72):

$$XP' = PH, \tag{5.84a}$$

$$XH' = PH + 2PR + HR, \tag{5.84b}$$

$$XR' = HR, \tag{5.84c}$$

with

$$X = 1 - P^2 - H^2 - R^2. \tag{5.85}$$

If $R = 0$, (5.84c) informs us that $R' = 0$, so we obtain from (5.84a) and (5.84b) $P' = H' = 1/2$. Thus, with $R(0) = 0$ the population reaches the equilibrium $P = H = 1/2$ in one generation. $H(0) = 0$ leads to extinction in two generations. Hence, we assume $R(0) > 0$ and $H(0) > 0$, whence $R(t) > 0$ and $H(t) > 0$. Dividing (5.84a) by (5.84c) immediately shows that

$$P(t)/R(t) = P(0)/R(0) \equiv k. \tag{5.86}$$

We derive an equation for $x = R/H$ by inserting (5.86) into the ratio of (5.84c) to (5.84b), with the result

$$x' = (1 + k + 2kx)^{-1}, \qquad 0 < x < \infty. \tag{5.87}$$

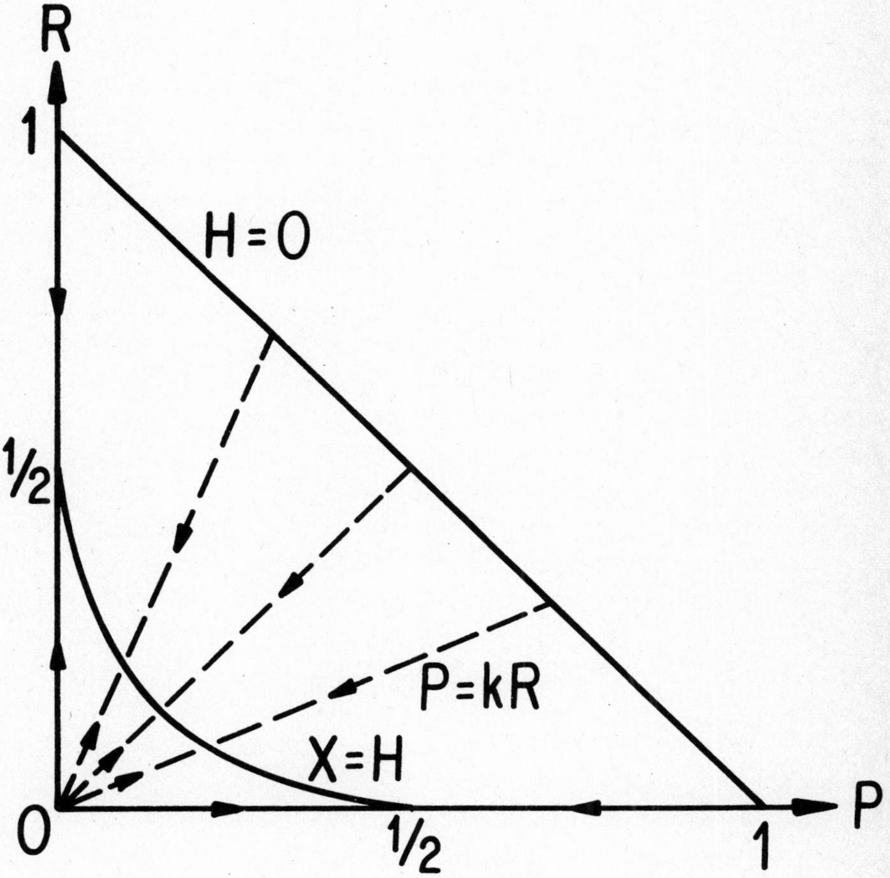

Fig. 5.5. The iteration of (5.84).

From Section 2.2 we see easily that (5.87) is Case 2(a) of the linear fractional transformation. Then (2.28) and (2.31) imply global oscillating convergence to

$$\hat{x} = (4k)^{-1}[-(1+k)+(1+10k+k^2)^{1/2}]$$

at the geometric rate μ^t, where

$$\mu = -(4k)^{-1}[1+6k+k^2-(1+k)(1+10k+k^2)^{1/2}].$$

Combining (5.86) with the definition of x, we get

$$\hat{R} = \hat{x}/[1+(1+k)\hat{x}].$$

We note from (5.84c) that the equilibria lie on the curve $X = H$, which, with the aid of (5.85), is revealed as the ellipse

$$2P^2 + 2PR + 2R^2 - 3P - 3R + 1 = 0.$$

We owe these results, displayed in Figure 5.5, to Finney (1952) and Karlin and Feldman (1968). Solutions were also presented by Cannings (1968) and Falk and Li (1969).

5.7 *Problems*

5.1. Consider a diallelic plant population with the usual notation, viz., $p = p_1$, $q = 1-p$, $P = P_{11}$, $Q = P_{12}$, $R = P_{22}$. Assume all AA plants in the infinitely large population are fertilized at random, and the allele a causes selfing, so that only a fraction α of Aa plants and a proportion γ of aa plants is pollinated at random. To allow for the possibility that selfing plants make a smaller contribution of randomly-fertilizing pollen than nonselfers, let κ and λ be the factors for Aa and aa selfers, respectively, by which this contribution is reduced compared to to cross-pollinators. Derive the recursion relations for P, Q, R. Show that

$$2T\Delta q = P(\mu Q + \nu R) + (\alpha\nu - \gamma\mu)QR,$$

where $\mu = (1-\alpha)\kappa$, $\nu = (1-\gamma)\lambda$, and

$$T = P + 2(\alpha+\mu)Q + (\gamma+\nu)R.$$

Deduce that if the selfing gene, a, is initially rare, it is established in the population (i.e., q increases) if and only if at least one of the selfing genotypes contributes to cross-fertilization. Notice that if $\alpha\nu > \gamma\mu$, the cross-pollinating allele is eventually lost (Nagylaki, 1976a).

5.2. Solve the difference equations

(a) $z' - \lambda z = \alpha t$, $z(0) = z_0$;

(b) $z'' - \lambda z = \mu^t$, $z(0) = z_0$, $z(1) = z_1$.

5.3. Show that the unique equilibrium of (5.37) satisfies $\hat{P} > p^2$ for $0 < p < 1$ and $0 \leq \alpha, \beta < 1$, unless $\alpha = \beta = 0$.

5.4. Derive (5.55a).

5.5. Derive (5.59), (5.67), and (5.70).

6. MIGRATION AND SELECTION

Since many natural populations are divided into subpopulations, it is important to investigate the effects of the joint action of selection and migration. We shall suppose that the population occupies a number of distinct niches, each of which has its own selection pattern. The niches may be defined by any pertinent set of environmental variables. If a population is distributed in clusters, this scheme provides a reasonable model for geographical variation. The study of the spatial variation of gene frequencies, however, requires difficult detailed analysis. In this chapter, we shall confine ourselves to the easier problem of the maintenance of genetic diversity. For time-dependent environments, we examined this question in Section 4.5 by introducing the concept of protected polymorphism. We shall use the same approach here to expound multi-niche selection.

We shall restrict our treatment to a single autosomal locus, and assume that the sexes need not be distinguished. Only genotype-independent migration will be treated. Selection operates entirely through viability differences. Most of our examples will concern the diallelic case. In Section 6.1, we shall give a complete analysis of the island model. The general formulation and analysis of Section 6.2 will be applied to the Levene (1953) model and to the two-niche situation in Sections 6.3 and 6.4.

6.1 *The Island Model*

We designate the frequencies of the allele A_i in zygotes and adults on our figurative island by p_i and p_i^*. The corresponding population sizes are N and N^*. We assume that a proportion α of adults is removed from the island either by mortality or emigration, and a fraction β of migrants with constant allelic frequencies $\overline{p}_i$ is added. If the migration mechanism is exchange with a metaphoric continent, its population must exceed that of the island sufficiently to allow us to neglect the change in $\overline{p}_i$. Clearly, the number of A_i genes in zygotes in the next generation is given by

$$n_i' = [(1-\alpha)p_i^* + \beta\overline{p}_i](2N^*),$$

whence the number of zygotes in the next generation reads

$$N' = \frac{1}{2}\sum_i n_i' = (1-\alpha+\beta)N^*.$$

Therefore,

$$p_i' = n_i'/(2N') = (1-m)p_i^* + m\bar{p}_i, \tag{6.1}$$

where

$$m = \beta/(1-\alpha+\beta).$$

Evidently, m is the fraction of zygotes with immigrant parents. If the influx and efflux of individuals are equal, then $\alpha = \beta = m$, the proportion of adults replaced per generation. We owe this special case of the model to Haldane (1930) and Wright (1931). Since efflux does not alter gene frequencies, the fact that, given the above reinterpretation, their formulation always applies is hardly surprising.

As explained in Section 2.3, (6.1) is a particular case of the mutation-selection model (4.95), (4.97).

To complete the model, we posit random mating after migration on the island. The only effect of reproduction here is to return the population to Hardy-Weinberg proportions. Then

$$p_i^* = p_i w_i/\bar{w}, \tag{6.2}$$

the fitnesses being specified by the Hardy-Weinberg formulae (4.25b). It is clear from (6.1) that no allele carried to the island by immigrants can be lost. Therefore, we shall focus our attention on the diallelic case with $\bar{p} = 0$, i.e., the immigrants are all aa.

Combining (6.1) and (6.2) we obtain the transformation

$$p' = (1-m)pw_1/\bar{w} \equiv f(p). \tag{6.3}$$

Since $f(1) = 1-m < 1$, for $m > 0$, in agreement with the remark in the previous paragraph, a cannot be lost. As $p \to 0$, (6.3) yields

$$f(p) \sim [(1-m)w_{12}/w_{22}]p. \tag{6.4}$$

The allele A will be protected if the above bracket exceeds unity, i.e., if

$$m < 1 - \frac{w_{22}}{w_{12}}. \tag{6.5}$$

Obviously, $p = 0$ is an equilibrium. If there is no other equilibrium, then $f(1) < 1$ implies $p' = f(p) < p$ for $0 < p \leq 1$, so A will be lost.

We shall now present a complete analysis of (6.3). The equilibria with $p \neq 0$ satisfy the quadratic

$$\overline{w} = (1-m)w_1. \tag{6.6}$$

Since we are interested in the maintenance of genetic variability by migration and selection, we suppose AA is favored over the immigrant genotype, aa, and, for simplicity, exclude over- and underdominance in the remainder of this section. Then we may parametrize the fitnesses by

$$w_{11} = 1+s, \qquad w_{12} = 1+hs, \qquad w_{22} = 1-s, \tag{6.7}$$

where s, $0 < s \leq 1$, represents the intensity of selection, and h, $-1 \leq h \leq 1$, determines the degree of dominance. We take $0 < m < 1$.

From (4.25b) and (6.7) we derive

$$w_1 = 1+hs+s(1-h)p, \tag{6.8a}$$

$$\overline{w} = 1-s+2s(1+h)p-2hsp^2. \tag{6.8b}$$

Substituting (6.8) into (6.6) produces the quadratic equilibrium equation

$$2hp^2-[1+3h+m(1-h)]p+[1+h(1-m)-\nu] = 0,$$

where $\nu = m/s \geq m$. Therefore, the equilibria read

$$p_{\pm} = (4h)^{-1}\{1+3h+m(1-h)\pm[(1-h)^2(1+m)^2+8h(\nu+m)]^{1/2}\}. \tag{6.9}$$

As $h \to 0$, (6.9) gives the correct limit,

$$p_- = (1-\nu)/(1+m), \qquad h = 0,$$

for the case without dominance. The equilibria depend on three parameters, h, m, and ν, the selection intensity s being absorbed into the migration-selection ratio ν. Of course, we require that p_+ and p_- be real and $0 \leq p_{\pm} \leq 1$. To find out what parameter ranges yield acceptable equilibria, we impose these conditions on (6.9). After some elementary but tedious calculation, the following results emerge (Problem 6.1).

Define $\mu = \nu^{-1} = s/m \leq m^{-1}$, the strength of selection relative to that of migration, and

$$h_0 = -(1+m)/(3-m), \tag{6.10}$$

$$\mu_1 = [1+h(1-m)]^{-1}, \tag{6.11a}$$

$$\mu_2 = -8h[(1-h)^2(1+m)^2 + 8hm]^{-1}. \tag{6.11b}$$

Observe that, as m increases from 0 to 1, h_0 decreases from $-1/3$ to -1. Hence, for $0 < m < 1$, if there is no dominance ($h = 0$) or A is dominant ($h = 1$), then $h > h_0$; if A is recessive ($h = -1$), then $h < h_0$. We fix h and m, and study p_+ and p_- as functions of μ, displaying the results in Figure 6.1. We remark that

$$p_-(\mu_2) = (4h)^{-1}[1+3h+m(1-h)], \tag{6.12a}$$

$$p_-(m^{-1}) = (4h)^{-1}\{1+3h+m(1-h)-[(1-h)^2(1+m)^2+16hm]^{1/2}\}. \tag{6.12b}$$

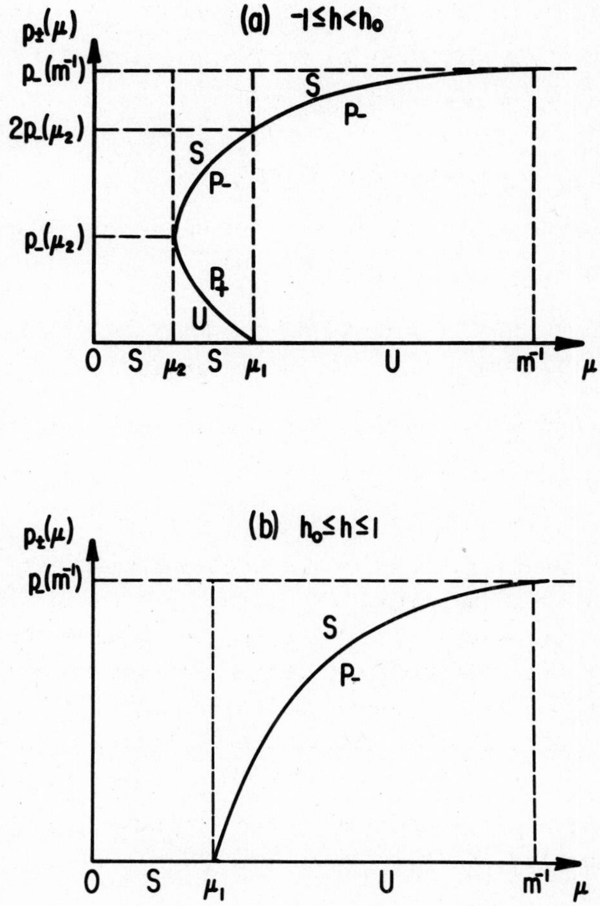

Fig. 6.1. The equilibrium and stability structure of (6.3).

For weak selection and migration, $s, m \ll 1$, in (6.9) to (6.12) the correct approximation is obtained by setting $m = 0$ wherever it appears explicitly.

Our next task is to justify the stability behavior indicated in Figure 6.1. There are three cases.

1. $\mu \leq \mu_2$ and $-1 \leq h < h_0$, or $\mu \leq \mu_1$ and $h_0 \leq h \leq 1$:

   In these two situations, since there is no internal equilibrium, by the comment below (6.5), $p(t) \to 0$ as $t \to \infty$, i.e., A is ultimately lost. (In the degenerate case $\mu = \mu_2$ and $-1 \leq h < h_0$, $p_+(\mu_2) = p_-(\mu_2)$ is an internal equilibrium, but if $p(0) < p_\pm(\mu_2)$, then $p(t) \to 0$; if $p(0) > p_\pm(\mu_2)$, then $p(t) \to p_\pm(\mu_2)$, and a slight perturbation to $p < p_\pm(\mu_2)$ will cause loss of A.)

2. $\mu_2 < \mu \leq \mu_1$ and $-1 \leq h < h_0$:

   We must distinguish two subcases.

   (a) $p(0) < p_+$:

   From (6.3) and (6.8) we find

$$\Delta p = 2hsp(p - p_+)(p - p_-)/\bar{w}. \tag{6.13}$$

   Therefore, if $0 < p < p_+$, $\Delta p < 0$ (recall that $h_0 < -1/3$), whence $p(t) \to 0$.

   (b) $p(0) > p_+$:

   We put $x = p - p_-$, and obtain from (6.13) $x' = \lambda(p)x$ with

$$\lambda(p) = 1 + \frac{2hsp(p - p_+)}{\bar{w}}. \tag{6.14}$$

   For $p > p_+$, $\lambda(p) < 1$, and, recalling (6.8b), we have

$$\lambda(p) \geq 1 + \frac{2hsp^2}{\bar{w}} = \frac{1 - s + 2s(1 + h)p}{\bar{w}} \geq 0.$$

   It follows from the general result in Section 4.2 that $p(t) \to p_-$ without oscillation.

3. $\mu > \mu_1$:

   Inserting (6.7) into (6.5) and recollecting (6.11a) tells us that, as is obvious from a glance at Figure 6.1, A is protected. To prove global convergence to p_-, we separate positive and negative values of h.

(a) $-1 \le h < 0$:

Direct algebra demonstrates that $p_+ < 0$, allowing us to draw the conclusion $\lambda(p) < 1$ from (6.14). Let us establish that $\lambda(p) \ge 0$. From (6.14) we obtain

$$\bar{w}\lambda(p) = 1-s+2s(1+h)p-2hspp_+. \tag{6.15}$$

Now, $hp_+ > 0$, and (6.9) gives

$$4hp_+ = 1+3h+m(1-h)+[(1-h)^2(1+m)^2+8h(\nu+m)]^{1/2}. \tag{6.16}$$

Since $h < 0$, (6.16) is obviously a monotone decreasing function of ν for fixed h and m. Remembering that $\nu \ge m$, we derive an upper bound for $4hp_+$ by substituting $\nu = m$ into (6.16):

$$4hp_+ \le 1+3h+m(1-h)+[(1-h)^2(1+m)^2+16hm]^{1/2}. \tag{6.17}$$

But, as is verified by squaring (6.18), for $m < 1$

$$[(1-h)^2(1+m)^2+16hm]^{1/2} < 3-m+h(1+m). \tag{6.18}$$

Inserting (6.18) into (6.17), we find $hp_+ < 1+h$. Using this in (6.15) gives the desired result,

$$\bar{w}\lambda(p) > 1-s \ge 0,$$

thereby proving nonoscillating global convergence.

(b) $0 < h \le 1$:

Straightforward algebra now shows that $p_+ > 1$, which permits us to infer that $\lambda(p) < 1$ from (6.14). With $h > 0$, (6.16) is evidently monotone increasing in ν for fixed h and m. Since $\nu < \mu_1^{-1}$, we get an upper bound on $4hp_+$ by substituting $\nu = \mu_1^{-1}$ into (6.16):

$$2hp_+ < 1+3h+m(1-h).$$

Inserting this into (6.15) leads to

$$\bar{w}\lambda(p) > 1-s+sp(1-h)(1-m) \ge 0,$$

again establishing that $p(t) \to p_-$ without oscillation.

(c) $h = 0$:

Since $2hp_+ \to 1+m$ as $h \to 0$, the conclusion for no dominance follows as a limit of either Case (a) or Case (b). It is also easy to analyze this situation *ab initio* (Problem 6.2).

For the much easier analysis with weak migration and selection, see Haldane (1930) and Nagylaki (1975b).

Several qualitative conclusions can be drawn from Figure 6.1. For the existence of a polymorphism, the intensity of selection compared to that of migration, μ, must exceed a critical value (μ_1 or μ_2), which is usually of the order of unity. For most parameter values, A is absent or has a high frequency; it is rare only if $h_0 \leq h \leq 1$ and μ slightly exceeds μ_1. If $-1 \leq h < h_0$, i.e., A is partially or completely recessive, as conditions for maintaining A become unfavorable in the sense that $\mu \to \mu_2+$, the equilibrium gene frequency $p_-(\mu) \to p_-(\mu_2) > 0$. One might naively have expected A to disappear, $p_-(\mu) \to 0$, as indeed it does for $h_0 \leq h \leq 1$. In sum, the comments of this paragraph suggest that, unlike mutation and selection, migration and selection may not maintain alleles at low frequencies at many loci, thereby possibly limiting response to environmental change.

6.2 *General Analysis*

We assume that the population occupies a number of distinct niches, each of which has its own selection scheme. To describe the migration among these colonies, we employ the forward and backward migration patterns first introduced by Malécot (1948) for the study of migration and random drift in the absence of selection. The forward and backward migration matrices (Bodmer and Cavalli-Sforza, 1968), $\tilde{M}$ and M, were first applied to the migration-selection problem by Bulmer (1972). Their elements, $\tilde{m}_{ij}$ and m_{ij}, may be functions of our sundry variables and parameters. We suppose that adults migrate, discussing alternative hypotheses below, and define $\tilde{m}_{ij}$ and m_{ji} as follows. The probability that an adult in niche i migrates to niche j is $\tilde{m}_{ij}$. Since every niche is assumed to have a very large population, $\tilde{m}_{ij}$ represents the fraction of adults migrating from niche i to niche j in each generation. We designate by m_{ji} the probability that an adult in niche j migrated from niche i. The migration matrices satisfy the normalization conditions

$$\sum_j \tilde{m}_{ij} = 1, \qquad \sum_i m_{ji} = 1. \tag{6.19}$$

The life cycle is selection (possibly including population regulation), migration, population regulation (if present at this stage), and reproduction. If it is the zygotic rather than the mating adult population that is regulated, we may interchange the last two steps without modifying our equations. Thus, the population may be controlled once, twice, or not at all in one generation. The proportions of zygotes, pre-migration adults, post-migration adults, and post-regulation adults in niche i are c_i, c_i^*, c_i^{**}, and c_i'. We denote the frequency of the allele A_j in niche i at the corresponding stages by $p_{i,j}$, $p_{i,j}^*$, $p_{i,j}'$, and $p_{i,j}'$. We summarize this information in the following scheme.

$$\text{Zygote} \xrightarrow{\text{selection}} \text{Adult} \xrightarrow{\text{migration}} \text{Adult} \xrightarrow{\text{regulation}} \text{Adult} \xrightarrow{\text{reproduction}} \text{Zygote}$$

$$c_i p_{i,j} \qquad c_i^* p_{i,j}^* \qquad c_i^{**} p_{i,j}' \qquad c_i' p_{i,j}' \qquad c_i' p_{i,j}'$$

According to this formulation, the only effect of reproduction is to return each subpopulation to Hardy-Weinberg proportions through random mating in each niche.

Since no individuals are lost in migration, we have

$$c_j^{**} = \sum_i c_i^* \tilde{m}_{ij}, \tag{6.20a}$$

$$c_i^* = \sum_j c_j^{**} m_{ji}. \tag{6.20b}$$

Expressing the joint probability that an adult is in niche i and migrates to niche j in terms of the prospective and retrospective conditional probabilities, $\tilde{m}_{ij}$ and m_{ji}, yields

$$c_i^* \tilde{m}_{ij} = c_j^{**} m_{ji}. \tag{6.21}$$

The appropriate summations and use of (6.19) in (6.21) recapture (6.20). Inserting (6.20a) into (6.21), we deduce the desired connection between the forward and backward migration matrices:

$$m_{ji} = c_i^* \tilde{m}_{ij} \Big/ \sum_k c_k^* \tilde{m}_{kj}. \tag{6.22}$$

We designate the fitness of $A_j A_k$ individuals in niche i by $w_{i,jk}$. Therefore,

$$p_{i,j}^* = p_{i,j} w_{i,j} / \bar{w}_i, \tag{6.23a}$$

where the allelic fitnesses and the mean fitness in niche i read

$$w_{i,j} = \sum_k w_{i,jk} p_{i,k}, \tag{6.23b}$$

$$\bar{w}_i = \sum_{jk} w_{i,jk} p_{i,j} p_{i,k}. \tag{6.23c}$$

The gene frequencies in the next generation may be calculated directly from the backward migration matrix:

$$p'_{i,j} = \sum_k p^*_{k,j} m_{ik}. \tag{6.24}$$

Observe that if we consider the flow of genes, it is the forward migration matrix that enters:

$$c_i^{**} p'_{i,j} = \sum_k c_k^* p^*_{k,j} \tilde{m}_{ki}. \tag{6.25}$$

Since (6.21) immediately reduces (6.25) to (6.24), we shall employ the simpler equation (6.24).

If we know the fitnesses, $w_{i,jk}$, and the distribution of immigrants into the panmictic subpopulations (called *demes*), M, then (6.23) and (6.24) completely determine the model. But if, instead of M, we are given the distribution of emigrants from the demes, $\tilde{M}$, to have a well-posed problem in (6.23) and (6.24), we must compute M from (6.22). This requires an *Ansatz* for $\underset{\sim}{c}^*$ in terms of $\underset{\sim}{c}$ and a hypothesis for the variation, if any, of $\underset{\sim}{c}$.

One must specify the adult distribution prior to migration, $\underset{\sim}{c}^*$, in accordance with the nature of selection. The two possible extreme assumptions were lucidly described by Dempster (1955). If the proportion of adults in every niche is fixed, a mode now usually called *soft selection* (Wallace, 1968), we have $c_i^* = c_i$. As Dempster (1955) observed, this should be a good approximation if the population is regulated within each niche. If it is the total population size that is controlled, Dempster pointed out that it is better to suppose that the fraction of zygotes in each niche is prescribed, and the fraction of adults is proportional to the mean fitness in the niche:

$$c_i^* = c_i \bar{w}_i / \bar{w}, \tag{6.26}$$

where

$$\bar{w} = \sum_i c_i \bar{w}_i \tag{6.27}$$

is the mean fitness of the population. Wallace (1968) termed (6.26) *hard selection*. Since plants compete for resources locally, soft selection must apply to them. For animals, the choice of scheme depends on the situation (Wallace, 1968).

Let us consider now the influence of the order of effects in our model. If we replace adult migration by gametic dispersion (as may be reasonable for some lower vascular plants), provided we denote the allelic frequencies after dispersion by $p_{i,j}$, no change in our formulation is required. This simple observation is disguised if allelic frequencies before dispersion are used (Bulmer, 1972) because, in contrast to our definition, they are not the same as the gene frequencies in zygotes.

If zygotes, rather than gametes, disperse, new formulation and analysis are required. We shall discuss this case at the end of this section.

The scheme (6.23), (6.24) includes the haploid case by virtue of the multiplicative fitness device mentioned in Section 4.1. Since selection is local, the substitution $w_{i,jk} = b_{i,j}b_{i,k}$ converts (6.23) to a haploid pattern.

Let us now derive sufficient conditions for the existence of a protected polymorphism in the diallelic case. We write p_i $(= p_{i,1})$ for the frequency of A in niche i, and seek sufficient conditions for the instability of the trivial equilibria $p_i = 0$ and $p_i = 1$. These guarantee that both alleles will increase in frequency when rare. We define the ratios of the homozygote fitnesses in niche i to that of the heterozygotes by

$$u_i = w_{i,11} / w_{i,12}, \qquad v_i = w_{i,22} / w_{i,12}. \tag{6.28}$$

We posit that u_i and v_i are constant. With frequency-dependent selection, $u_i = u_i(p_i)$ and $v_i = v_i(p_i)$, the protection conditions would involve only $u_i(1)$ and $v_i(0)$.

We commence with soft selection, $c_i^* = c_i$. Then (6.22) becomes

$$m_{ji} = c_i \tilde{m}_{ij} / \sum_k c_k \tilde{m}_{kj}. \tag{6.29}$$

We suppose that M is constant. This is equivalent to constant $\tilde{M}$ if $\underset{\sim}{c}$ is constant. Without post-migration population regulation, $\underset{\sim}{c}$ will generally depend on time. Indeed, in that case from (6.20a) we have

$$\underset{\sim}{c}' = \underset{\sim}{c}^{**} = L\underset{\sim}{c},$$

whence

$$\underset{\sim}{c}(t) = L^t \underset{\sim}{c}(0),$$

where L is the transpose of $\tilde{M}$. Therefore, our results are more likely to apply with population control after migration: $\underset{\sim}{c}' = \underset{\sim}{c}$. In some cases, $M = \tilde{M}$. One example is the random outbreeding and site homing model (Deakin, 1966; Maynard Smith, 1966, 1970; Christiansen, 1974, 1975), of which the Levene (1953) model, discussed in the next section, is a special case. If $\tilde{M}$ is symmetric and the demes are of equal size, from (6.19) and (6.29) we find directly that $M = \tilde{M}$.

To derive a sufficient condition for the protection of A, we linearize (6.23) near $\underset{\sim}{p} = \underset{\sim}{0}$. Defining the diagonal matrix D by $d_{ii} = v_i^{-1}$ and putting $Q = MD$, from (6.23), (6.24), and (6.28) we obtain the linearized recursion relation

$$\underset{\sim}{p}' = Q\underset{\sim}{p} \quad \text{with} \quad q_{ij} = m_{ij}/v_j. \tag{6.30}$$

Recalling (5.15), we conclude that if the absolute value of at least one eigenvalue of Q exceeds unity, then $\underset{\sim}{p} = \underset{\sim}{0}$ is unstable (Bulmer, 1972).

It is obvious from (6.28) and (6.30) that heterozygote lethality precludes protection. With viable heterozygotes, the allele A cannot be lost by a jump of the exact system (6.23), (6.24) from $p_i > 0$ to $p_i = 0$. Now, Q may have some eigenvalues less than 1 in absolute value. The corresponding eigenspace extends to a stable manifold of (6.23), (6.24), from which $\underset{\sim}{p}(t)$ would converge to $\underset{\sim}{0}$ were it not for the biological certainty of small perturbations that would move the population off this manifold. We must still ascertain, however, that the population does not repeatedly approach the stable eigenspace near $\underset{\sim}{p} = \underset{\sim}{0}$ arbitrarily closely. For if it did, say by virtue of suitable cycling, random drift would eventually cause loss of A.

We must require that individuals in every niche be able eventually to reach every other niche. Therefore, we suppose henceforth that there do not exist complementary sets of integers I and J such that $m_{ij} = 0$ for i in I and j in J, i.e., the matrix M is *irreducible*. So, Q is irreducible. Since Q is clearly nonnegative ($q_{ij} \geq 0$ for all i, j), Frobenius' theorem (Gantmacher, 1959, vol. II, p. 53) informs us that it has a real, positive, nondegenerate (= simple) eigenvalue, λ_0, (designated the *maximal* eigenvalue), not exceeded by any other eigenvalue in modulus, to which corresponds a positive eigenvector (i.e., all components of this *maximal* eigenvector may be chosen as positive). By employing the general decomposition (5.13) and the fact that, if a subspace is invariant with respect to an operator, then its orthogonal complement is invariant with respect to the adjoint operator (Gantmacher, vol. I, p. 266), one can prove that the positive maximal left eigenvector of Q is orthogonal to every vector in the eigenspace complementary to the maximal (right) eigenvector. Therefore, there are no nonnegative vectors in that space, and consequently $\lambda_0 > 1$ does imply protection of A.

We see at once from (6.28) that (6.30) does not apply if a is lethal in the homozygous state, $w_{i,22} = 0$. But in this case, as $p_i \to 0$, $p_i^* \to 1/2$ from (6.23), as is obvious, and hence $p_i' \to 1/2$ from (6.24), again obviously. Therefore, A, being present in every individual, is protected.

The maximal eigenvalue λ_0 satisfies the inequality

$$\min_i \sum_j q_{ij} \le \lambda_0 \le \max_i \sum_j q_{ij}, \tag{6.31}$$

with equality if and only if all the row sums are the same (Gantmacher, vol. II, p. 63). Suppose aa is at least as fit as Aa in every niche and more fit in at least one, $v_i \ge 1$ for all i and $v_i > 1$ for some i. Evidently, $q_{ij} \le m_{ij}$; irreducibility of M means there is no j such that $m_{ij} = 0$ for all i. Therefore, the row sums in (6.31) are not all equal, and (6.19) and (6.31) yield

$$\lambda_0 < \sum_j m_{ij} = 1.$$

As expected, A is not protected. If Aa is at least as fit as aa in all niches and more fit in at least one, the above reasoning tells us that $\lambda_0 > 1$, i.e., A is protected. For positive M ($m_{ij} > 0$), these results were proved by a different method by Bulmer (1972). The generalization to a nonnegative backward migration matrix is not academic: many important examples such as nearest-neighbor migration involve zeroes in M.

Consider (6.23) and (6.24) in the absence of migration, $m_{ij} = \delta_{ij}$. It is easy to locate all the equilibria of this system. Karlin and McGregor (1972, 1972a) have shown that if 1 is not an eigenvalue of Q, then for sufficiently weak migration, in the neighborhood of each stable equilibrium of the uncoupled system, there exists exactly one equilibrium of the weakly coupled one, and it is stable. In the neighborhood of each unstable equilibrium of the system without migration, there exists at most one equilibrium of the system with weak migration; if it exists, such an equilibrium is unstable. Now, if Aa is favored over aa in at least one niche (it may be less fit than aa in the other niches), i.e., $v_i < 1$ for at least one i, then in the absence of migration the equilibrium $\underset{\sim}{p} = \underset{\sim}{0}$ is unstable. By the above principle, for sufficiently small migration, this equilibrium is still unstable. Therefore, A will remain in the population. Although still intuitively reasonable, this result is much less obvious than the two in the previous paragraph. For weak migration, the latter also follow directly from the general principle.

For a protected polymorphism we stipulate also that a be preserved. As $p_i \to 1$, we obtain instead of m_{ij}/v_j the matrix elements m_{ij}/u_j, and again require that the maximal eigenvalue exceed unity.

If there is complete dominance, the eigenvalue condition cannot be satisfied: e.g., if a is dominant, $v_i = 1$, so $q_{ij} = m_{ij}$, and (6.31) gives $\lambda_0 = 1$. We shall study this important case after investigating hard selection.

Christiansen (1974) has derived a sufficient condition for the protection of A which is equivalent to $\lambda_0 > 1$, but is easier to apply. Let $I^{(n)}$ and $Q^{(n)}$ designate the $n \times n$ unit matrix and the square matrix formed from the first n rows and columns of Q. We require

$$\det\left(I^{(n)} - Q^{(n)} \right) < 0 \tag{6.32}$$

for some n, $n = 1, 2, \ldots, N$, where N is the number of niches.

For further analysis, see Karlin (1976) and Karlin and Richter-Dyn (1976).

We turn now to the hard selection model. Substituting (6.26) into (6.22), we find

$$m_{ji} = c_i \bar{w}_i \tilde{m}_{ij} \Big/ \sum_k c_k \bar{w}_k \tilde{m}_{kj}. \tag{6.33}$$

Since $\underset{\sim}{c}^*$ depends on gene frequency, M is now quite unlikely to remain constant. It is much more probable that $\tilde{M}$ is constant. Provided we suppose that the population is regulated after migration, $\underset{\sim}{c}' = \underset{\sim}{c}$, we can evaluate M from (6.33). As $\underset{\sim}{p} \to 0$, (6.23), (6.24), and (6.33) yield the linearized matrix difference equation

$$\underset{\sim}{p}' = R\underset{\sim}{p}, \qquad \text{where} \qquad r_{ij} = c_j w_{j,12} \tilde{m}_{ji} \Big/ \sum_k c_k w_{k,22} \tilde{m}_{ki}. \tag{6.34}$$

A will be protected if the maximal eigenvalue of R exceeds unity (Christiansen, 1975). To protect a, we require the same condition with $w_{k,22}$ replaced by $w_{k,11}$ in (6.34).

We can compare soft and hard selection by assuming that the fitnesses $w_{i,jk}$, the forward migration matrices $\tilde{M}$, and the zygotic distributions $\underset{\sim}{c}$ are the same. Having no further use for (6.33) here, we employ (6.26) to define the constant backward migration matrix M for soft selection, and rewrite (6.34) in the form $R = D^{(1)} M D^{(2)}$, where the diagonal matrices $D^{(1)}$ and $D^{(2)}$ have the elements

$$d_{ii}^{(1)} = \sum_k d_k \tilde{m}_{ki} \Big/ \sum_k c_k w_{k,22} \tilde{m}_{ki}, \qquad d_{ii}^{(2)} = w_{i,12}. \tag{6.35}$$

But for any two matrices B and C, BC and CB have the same eigenvalues (see Problem 6.3). Consequently, the eigenvalues of R are identical to those of $S = MD^{(2)}D^{(1)}$. Since (6.35) gives $s_{ij} = m_{ij} / \nu_j$, where

$$\nu_j = \left(\sum_k c_k w_{k,22} \tilde{m}_{kj} \right) \left(w_{j,12} \sum_k c_k \tilde{m}_{kj} \right)^{-1}, \tag{6.36}$$

we obtain for hard selection just the soft selection matrix (6.30) with v_j replaced by ν_j. Recalling (6.28), we may rephrase this in terms of the substitution

$$w_{j,22} \to \sum_k c_k w_{k,22} \tilde{m}_{kj} \Big/ \sum_k c_k \tilde{m}_{kj}.$$

We shall see in Section 6.3 that even in the special case of the Levene (1953) model no general statement can be made regarding the relative stringency of the protection conditions for soft and hard selection.

Our discussion of the case of recessive lethal a for soft selection applies unaltered here.

If migration is sufficiently small, the principle of Karlin and McGregor (1972, 1972a) described above demonstrates that A is protected if Aa is fitter than aa in at least one niche, and it is not protected if Aa is less fit than aa in every niche.

We can derive the other results corresponding to those we obtained for soft selection by rewriting r_{ij} in (6.34) as

$$r_{ij} = m_{ij}^*/v_j, \qquad \text{where} \qquad m_{ij}^* = c_j w_{j,22} \tilde{m}_{ji} \Big/ \sum_k c_k w_{k,22} \tilde{m}_{ki}. \tag{6.37}$$

Naturally, M^* is the backward migration matrix computed from (6.33) in the limit $p_i \to 0$. Comparing (6.30) and (6.37), and noting that

$$\sum_j m_{ij}^* = 1,$$

we deduce, precisely as for soft selection,

(i) if aa is at least as fit as Aa in every niche and more fit in at least one, A is not protected;

(ii) if aa and Aa are interchanged in (i), A is protected;

(iii) if a is dominant, the maximal eigenvalue of R is 1, so protection must be analyzed *ab initio*.

The most interesting feature of hard selection, obvious by inspection of (6.34), (6.36), or (6.37), is that, in contrast to soft selection, the process (including the protection conditions) can no longer be expressed in terms of the fitness ratios (6.28). Mathematically, this happens because the hard selection backward migration matrix, (6.33), depends on the fitnesses, while the one for soft selection, (6.29), does not. The biological reason is that with hard selection the fitnesses affect the deme sizes, and, since the gene frequencies in the demes are different, subpopulation numbers influence gene frequencies through migration. If (and only if) the heterozygote fitnesses are the same in all the niches, (6.34) simplifies to the result of Christiansen (1975),

$$r_{ij} = c_j \tilde{m}_{ji} \Big/ \sum_k c_k v_k \tilde{m}_{ki}. \tag{6.38}$$

A *Recessive*:

We have $w_{i,12} = w_{i,22}$ for all i. For soft selection, a will be protected if the maximal eigenvalue of the matrix with elements m_{ij}/u_j exceeds one. For hard selection, the pertinent matrix elements are

$$c_j w_{j,12} \tilde{m}_{ji} \Big/ \sum_k c_k w_{k,11} \tilde{m}_{ki}.$$

We wish to deduce sufficient conditions for protection of A. We must now compute the effect of selection to second order as $p_i \to 0$: (6.23) and (6.28) yield

$$p_i^* = p_i + (u_i - 1)p_i^2 + O(p_i^3). \tag{6.39}$$

Soft and hard selection require separate analyses.

With soft selection, we assume M is constant. In view of (6.19) and (6.31), M has the maximal eigenvalue 1 with the maximal eigenvector

$$\underset{\sim}{V} = \begin{pmatrix} 1 \\ 1 \\ \vdots \\ 1 \end{pmatrix}.$$

Let us suppose that all the other eigenvalues are (strictly) less than 1 in absolute value. By Perron's theorem (Gantmacher, Vol. II, p. 53), for this restriction to hold, it is sufficient, but not necessary, that M be positive. Substituting (6.39) into (6.24) produces

$$\underset{\sim}{p}' = M\underset{\sim}{p} + \underset{\sim}{f}(\underset{\sim}{p}) + O(|\underset{\sim}{p}|^3), \tag{6.40a}$$

where

$$f_i(\underset{\sim}{p}) = \sum_j m_{ij}(u_j - 1)p_j^2. \tag{6.40b}$$

For $|\underset{\sim}{p}| << 1$ and long times, we may invoke (5.15) to infer from (6.40) that $\underset{\sim}{p}(t) \approx x(t)\underset{\sim}{V}$, where $x(t)$ is the component of $\underset{\sim}{p}(t)$ along $\underset{\sim}{V}$. The component of $\underset{\sim}{p}(t)$ in the eigenspace of M complementary to $\underset{\sim}{V}$ decreases geometrically according to the dominant linear part of (6.40a). Since this observation holds to first order, we may replace $\underset{\sim}{p}$ by $x\underset{\sim}{V}$ in the higher order terms in (6.40a):

$$\underset{\sim}{p}' = M\underset{\sim}{p} + \underset{\sim}{f}(x\underset{\sim}{V}) + O(x^3). \tag{6.41}$$

By Frobenius' theorem (Gantmacher, Vol. II, p. 53), we may choose the left eigenvector, $\underset{\sim}{U}$, of M corresponding the eigenvalue 1 to be positive, and we also take $\underset{\sim}{U}^T\underset{\sim}{V} = 1$, where the superscript T signifies transposition. A will be protected if $x(t) = \underset{\sim}{U}^T\underset{\sim}{p}(t)$, the component of $\underset{\sim}{p}$ along $\underset{\sim}{V}$, increases (note that $\underset{\sim}{U}$ is orthogonal to the complement of $\underset{\sim}{V}$). Taking the scalar product of (6.41) with $\underset{\sim}{U}$ and using the fact that $\underset{\sim}{U}^TM = \underset{\sim}{U}^T$, we find

$$x' = x + \underset{\sim}{U}^T\underset{\sim}{f}(x\underset{\sim}{V}) + O(x^3). \tag{6.42}$$

The positivity of x allows us to derive from (6.42) the sufficient condition $\underset{\sim}{U}^T\underset{\sim}{f}(x\underset{\sim}{V}) > 0$ for protection of A. With the aid of (6.19), (6.40b), $\underset{\sim}{U}^{\tilde{T}}M = \underset{\sim}{\tilde{U}}^T$, and the normalization

$$\sum_i U_i = 1 \tag{6.43}$$

of $\underset{\sim}{U}$, we obtain the explicit criterion

$$\underset{\sim}{U}^T\underset{\sim}{u} = \sum_i U_iu_i > 1. \tag{6.44}$$

This condition depends only on the fitness ratios u_i and the maximal left eigenvector $\underset{\sim}{U}$ of the backward migration matrix M. The deme sizes c_i affect (6.44) only to the extent they influence $\underset{\sim}{U}$.

From (6.43) and (6.44) we can easily see that if AA is at least as fit as Aa in every niche and more fit in at least one, then A is protected. If AA and Aa are interchanged in the previous statement, A is not protected.

Christiansen (1974) obtained the condition

$$\underset{\sim}{c}^TM\underset{\sim}{u} > 1 \tag{6.45}$$

as the criterion for protection with a constant zygotic distribution $\underset{\sim}{c}$. This condition agrees with (6.44) if and only if $\underset{\sim}{c}^TM = \underset{\sim}{U}^T$. This is true if $\underset{\sim}{c} = \underset{\sim}{U}$, and, should $\det M = 0$, may be fortuitously valid for particular vectors $\underset{\sim}{c}$. From (6.20b) we conclude at once that $\underset{\sim}{c}^TM = \underset{\sim}{c}^T$

if and only if migration does not alter the subpopulation sizes. This holds for random outbreeding and site homing (Deakin, 1966; Maynard Smith, 1966, 1970; Christiansen, 1974) and its special case, the Levene (1953) model, but is not a general phenomenon. For example, if M is symmetric, evidently $\underset{\sim}{U} = N^{-1}\underset{\sim}{V}$, so (6.19) reduces (6.44) to

$$N^{-1} \sum_j u_j > 1. \tag{6.46}$$

This very simple condition on the unweighted mean of the fitness ratios agrees with (6.45) if the N subpopulations are of equal size, in which case migration does not change the deme sizes, but generally not otherwise.

Christiansen's condition, (6.45), is generally false because he replaced $\underset{\sim}{p}(t)$ by $x(t)\underset{\sim}{V}$ in the linear term ($M\underset{\sim}{p}$) on the right-hand side of (6.40a). The resulting equation is correct only to first order in $\underset{\sim}{p}$, and protection is due to the quadratic terms.

Let us consider now the case of hard selection. As before, we suppose $\tilde{M}$ and $\underset{\sim}{c}$ are constant. The selection equation (6.39) still applies, of course, but we must now calculate M from (6.33). From (6.23c) we find

$$\overline{w}_i = w_{i,22}[1 + (u_i - 1)p_i^2],$$

whence (6.33) immediately informs us that

$$M = M^* + O(|\underset{\sim}{p}|^2), \tag{6.47}$$

where M^* is the backward migration matrix in (6.37). Inserting (6.39) and (6.47) into (6.24) gives (6.40) with M replaced by M^*. Assuming that all eigenvalues of M^* other than 1 are less than 1 in modulus, from (6.44) we have the protection condition

$$\underset{\sim}{Y}^T \underset{\sim}{u} = \sum_i Y_i u_i > 1, \tag{6.48}$$

where $\underset{\sim}{Y}$ is the maximal left eigenvector of M^* and

$$\sum_i Y_i = 1. \tag{6.49}$$

As expected, the obvious conclusions for *AA* favored over *Aa* and *vice versa* follow as for soft selection. Again, the hard selection result depends on fitnesses, not just on fitness ratios. We shall see that even in the Levene model one can draw no general conclusion concerning the relative stringency of (6.44) and (6.48). If M^* is symmetric, (6.48) simplifies to (6.46).

Juvenile Dispersion:

For some marine organisms and for plants with seeds carried by wind, water, or animals, migration is best modeled by zygotic (or, at least, juvenile) dispersion. If the frequency of A_j in niche i immediately after meiosis is $p_{i,j}$, giving zygotic frequencies $p_{i,j}p_{i,k}$ just after conception, the post-migration zygotic proportions read

$$P_{i,jk} = \sum_l p_{l,j}p_{l,k}m_{il}, \tag{6.50}$$

where M is now the zygotic backward migration matrix. Including selection yields the recursion relations

$$p'_{i,j} = \sum_k w_{i,jk}P_{i,jk} \Big/ \sum_{kl} w_{i,kl}P_{i,kl}. \tag{6.51}$$

The model described by (6.50) and (6.51) is quite different from the one specified by (6.23) and (6.24).

With the zygotic distribution $\underset{\sim}{c}$ regulated to constancy, we may assume M is constant regardless of the nature of selection. Then it is easy to derive the criteria for a protected polymorphism in the diallelic case. For protection of A, we linearize (6.50) and (6.51) as p_i, the frequency of A in niche i, tends to zero. An easy calculation yields

$$\underset{\sim}{p}' = T\underset{\sim}{p}, \qquad \text{with} \qquad t_{ij} = m_{ij}/v_i. \tag{6.52}$$

Since $T = DM$ has the same eigenvalues as MD (see Problem 6.3), comparing (6.30) and (6.52), we see at once that the sufficient conditions for protection are the same as for adult migration with soft selection (Christiansen, 1975). The discussion between (6.30) and (6.32) applies without essential change.

It remains only to consider the case of recessive A. Expanding (6.50) and (6.51) to second order in $|\underset{\sim}{p}|$, we find

$$\underset{\sim}{p}' = M\underset{\sim}{p} + \underset{\sim}{g}(\underset{\sim}{p}) + O(|\underset{\sim}{p}|^3), \tag{6.53a}$$

where

$$g_i(\underset{\sim}{p}) = p_i \sum_j m_{ij}p_j - \left(\sum_j m_{ij}p_j\right)^2 + (u_i - 1)\sum_j m_{ij}p_j^2. \tag{6.53b}$$

Obviously, (6.40a) and (6.53a) have the same form. Therefore, A will be protected if $\underset{\sim}{U}^T\underset{\sim}{g}(x\underset{\sim}{V}) > 0$, where the maximal left eigenvector, $\underset{\sim}{U}$, of M is again normalized as in (6.43). By a trite reduction employing (6.19) and (6.43) we obtain again the criterion (6.44). Thus, in all cases the sufficient conditions for protection with juvenile dispersion are the same as with adult migration.

6.3 *The Levene Model*

Since we are primarily interested in the sufficient conditions for a protected polymorphism, the result at the end of the last section permits us to restrict our discussion to adult migration. The basic assumption of the Levene (1953) model is that migration is random. More precisely, individuals disperse independently of their niche of origin: $\tilde{m}_{ij} = \mu_j$, for some constants μ_j. We treat soft and hard selection separately.

For soft selection, (6.22) directly yields $m_{ji} = c_i$. If the demes are regulated to constant proportions, this is all we need. But the proportions remain constant even without regulation, for (6.20a) gives $c_j' = c_j^{**} = \mu_j$. Thus, we have the standard interpretation, $c_i = \mu_i$.

We proceed to demonstrate that the geometric mean, w^*, of the average fitnesses in the various niches is nondecreasing, the change in w^* being zero only at equilibrium. Such a property has been proved only in the Levene model. From (6.24) we observe that, not surprisingly, after one generation the gene frequencies in zygotes are the same in all the niches. Therefore, we may simplify our notation by letting p_i represent the frequency of A_i in zygotes. Inserting (6.23a) into (6.24), we find

$$p_i' = p_i \sum_j c_j w_{j,i} / \overline{w}_j. \tag{6.54}$$

Since (6.23b) and (6.23c) may be rewritten as

$$w_{i,j} = \sum_k w_{i,jk} p_k, \tag{6.55a}$$

$$\overline{w}_i = \sum_{jk} w_{i,jk} p_j p_k, \tag{6.55b}$$

we can express (6.54) in the form

$$p_i' = \frac{p_i}{2} \sum_j \frac{c_j}{\overline{w}_j} \frac{\partial \overline{w}_j}{\partial p_i} = \frac{p_i}{2} \sum_j c_j \frac{\partial}{\partial p_i} \ln \overline{w}_j = \frac{p_i}{2} \frac{\partial}{\partial p_i} \ln w^*, \tag{6.56}$$

where

$$w^* = \prod_i \overline{w}_i^{c_i}. \tag{6.57}$$

We might hope that w^* is nondecreasing because (6.56) has the form of multiallelic selection in a single panmictic population. Proof is required, however: for random mating in a single population, the mean fitness is a homogeneous quadratic in $\underset{\sim}{p}$, which w^* manifestly is not.

From (6.55), (6.56), and (6.57) we can easily verify that

$$\sum_i p_i' = 1.$$

Making this explicit in (6.56) leads to

$$p_i' = p_i \frac{\partial w^*}{\partial p_i} \Big/ \sum_j p_j \frac{\partial w^*}{\partial p_j}. \tag{6.58}$$

We can approximate (6.57), and hence the orbit $\underset{\sim}{p}(t)$, as closely as desired by the replacement

$$w^* \to W = \prod_i \bar{w}_i^{\gamma_i},$$

with positive rational γ_i, provided we choose γ_i sufficiently close to c_i. There exists a positive integer l such that $F \equiv W^l$ is a homogeneous polynomial in p with nonnegative coefficients; with arbitrary accuracy (6.58) becomes

$$p_i' = p_i \frac{\partial F}{\partial p_i} \Big/ \sum_j p_j \frac{\partial F}{\partial p_j}.$$

Now it follows directly from a result of Baum and Eagon (1967) that $F(\underset{\sim}{p})$ is nondecreasing along trajectories, and the change in F is zero only at equilibrium. The same obviously holds for W, and consequently for $w^*(\underset{\sim}{p})$.

The above argument deviates from that of Cannings (1971) only in replacing w^* by F to meet the restrictions of Baum's and Eagon's theorem. The existence of the nondecreasing function w^* implies that the system (6.54) must converge to some equilibrium point for all initial conditions. In particular, for the diallelic case discussed below, if there is a protected polymorphism, the gene frequency must converge to some stable internal equilibrium point.

Passing to two alleles, it is easy to deduce the criteria for protection directly from (6.54). But even this simple calculation is unnecessary. Notice that $q_{ij} = c_j/v_j$, independent of i, in (6.31). Therefore, the left- and right-hand sides of (6.31) are equal, and

$$\lambda_0 = \sum_i c_i/v_i.$$

So, A will be protected from disappearance if the harmonic mean of the fitness ratios v_i is less than one (Levene, 1953):

$$\tilde{v} \equiv \left(\sum_i c_i / v_i \right)^{-1} < 1. \tag{6.59}$$

Clearly, a will be protected if (6.59) holds with v_i replaced by u_i.

If A is recessive, $v_i = 1$, we can again derive the sufficient condition for its protection by expanding (6.54). Observing that $\underset{\sim}{c}$ is the maximal left eigenvector of M, however, we see at once from (6.44) that the arithmetic mean of the fitness ratios u_i must exceed unity (Prout, 1968):

$$\bar{u} \equiv \sum_i c_i u_i > 1. \tag{6.60}$$

Thus, a sufficient condition for a protected polymorphism reads

$$\tilde{u} < 1 < \bar{u}. \tag{6.61}$$

Equation (6.60) requires superiority of AA over Aa in the mean of the fitness ratios $w_{i,11}/w_{i,12}$. Now, $\bar{u} \geq \tilde{u}$, with equality if and only if all the u_i are the same (Problem 6.6). Therefore, fitness ratios u_i certainly exist for which (6.61) applies. Furthermore, (6.59) can be satisfied without heterozygote advantage in the mean of the v_i ($\bar{v} < 1$). Such superiority does imply (6.59).

We continue with hard selection, assuming the zygotic distribution $\underset{\sim}{c}$ is regulated to constancy. With $\tilde{m}_{ij} = \mu_j$, from (6.33) we find

$$m_{ji} = c_i \bar{w}_i / \bar{w}. \tag{6.62}$$

Therefore, as for soft selection, (6.24) informs us that after one generation the allelic frequencies in zygotes in all the niches are equal. Then we may again designate the zygotic frequency of A_i by p_i; $\bar{w}_i$ and $\bar{w}$ are given by (6.55b) and (6.27). Inserting (6.23a) and (6.62) into (6.24), we obtain

$$p_i' = p_i \bar{w}^{-1} \sum_k c_k w_{k,i}. \tag{6.63}$$

The arithmetic mean fitnesses of A_iA_j, A_i, and the entire population read

$$z_{ij} = \sum_k c_k w_{k,ij}, \tag{6.64a}$$

$$z_i = \sum_j z_{ij} p_j, \tag{6.64b}$$

$$\bar{z} = \sum_{ij} z_{ij} p_i p_j. \tag{6.64c}$$

Employing (6.55) and (6.64a) in (6.64b) and (6.64c) yields

$$z_i = \sum_k c_k w_{k,i}, \qquad \bar{z} = \bar{w},$$

with the aid of which (6.63) simplifies to

$$p_i' = p_i z_i / \bar{z}, \tag{6.65}$$

the classical selection equation for a single random mating population with fitnesses z_{ij}. For two alleles, this result is due to Dempster (1955).

It follows from Section 4.3 that the mean fitness $\bar{z}$ is nondecreasing, and the population converges to equilibrium point(s). With two alleles, Section 4.2 tells us that A is protected if and only if either

$$z_{12} > z_{22} \tag{6.66a}$$

or

$$z_{11} > z_{12} = z_{22}. \tag{6.66b}$$

Of course, (6.66b) means A is recessive. There is a protected polymorphism if and only if arithmetic mean overdominance applies:

$$z_{12} > z_{11}, z_{22}. \tag{6.67}$$

Let us compare the stringency of the conditions for a protected polymorphism with soft and hard selection. If there is complete dominance, the soft selection criterion is (6.61), while (6.67) rules out protection. In the absence of complete dominance, we desire to compare the requirements for protecting A, given by (6.59) for soft and (6.66a) for hard selection. From (6.28) and (6.64a) it is easy to see that if aa has the same fitness in every niche, then (6.59) and (6.66a) are the same. If the heterozygote fitness is the same in every niche, (6.66a) becomes

$$\bar{v} = \sum_i c_i v_i < 1. \tag{6.68}$$

But the arithmetic mean exceeds the harmonic mean unless all the fitness ratios are equal, in which case $\tilde{v} = \bar{v}$ (Problem 6.6). Hence, protection of A with hard selection implies its protection with soft selection, but not *vice versa*.

It is not true, however, that the hard selection criterion, (6.66a), is always more restrictive than the soft selection criterion, (6.59). Glancing at (6.59) and (6.66a), it is clear that the soft selection condition is more stringent than the hard one if and only if

$$\tilde{v} > z_{22}/z_{12},$$

which we rearrange in the form

$$\sum_i c_i w_{i,12} > \left(\sum_i c_i w_{i,22}\right)\left(\sum_i c_i w_{i,12}/w_{i,22}\right). \tag{6.69}$$

To give an example of protection with hard selection which does not imply protection with soft selection, we take $w_{i,12} = w_{i,22}^2$. Then (6.69) reduces to

$$\sum_i c_i w_{i,22}^2 > \left(\sum_i c_i w_{i,22}\right)^2. \tag{6.70}$$

Choosing $\mu = 2$ in the special case (4.47) of Jensen's inequality informs us that (6.70) holds as long as not all the fitnesses $w_{i,22}$ are the same.

6.4 *Two Diallelic Niches*

As in Section 6.3, we shall treat only adult migration. Since the hard selection protection criteria may be obtained from the soft selection ones by the replacement $M \to M^*$, the limiting backward migration matrix M^* being given by (6.37), we shall restrict ourselves to soft selection. We employ the notation of Section 6.2, setting $v_i = 1 - s_i$, and seek sufficient conditions for protection of A.

We shall consider the case of recessive A, $s_1 = s_2 = 0$, separately below. Following (6.31) we demonstrated that if $s_i \le 0$ for all i and $s_i < 0$ for some i, A is not protected. We proved also that $s_i \ge 0$ for all i and $s_i > 0$ for some i ensures preservation of A in the population. Therefore, we may assume $s_1 s_2 < 0$, i.e., neither selection coefficient vanishes, and s_1 and s_2 have the opposite sign. Putting

$$M = \begin{pmatrix} 1 - m_1 & m_1 \\ m_2 & 1 - m_2 \end{pmatrix}, \tag{6.71}$$

from (6.30) and (6.32) we find that A is protected if either

$$m_1 < s_1 \tag{6.72a}$$

or

$$\frac{m_1}{s_1} + \frac{m_2}{s_2} < 1. \qquad (6.72b)$$

Now, (6.72a) fails if $s_1 < 0$, although (6.72b) may not. If $s_1 > 0$ and $s_2 < 0$, (6.72a) is more severe than (6.72b). Therefore, we may confine ourselves to the weaker requirement (6.72b). The region of protection of A is displayed in Figure 6.2 for fixed m_i.

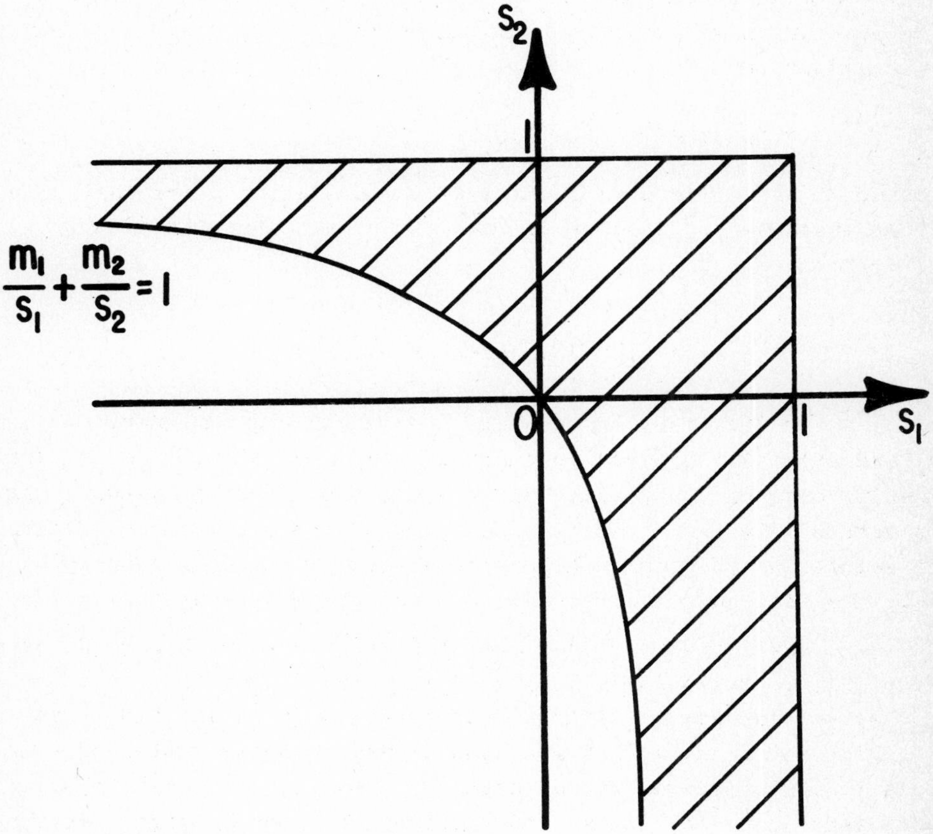

Fig. 6.2. The region of protection of A (hatched).

The backward migration matrix for random outbreeding and site homing (Deakin, 1966; Maynard Smith, 1966, 1970; Christiansen, 1974) with an arbitrary number of niches has the elements

$$\begin{aligned} m_{ii} &= 1 - k + kc_i, \\ m_{ij} &= kc_j, \qquad i \neq j. \end{aligned} \tag{6.73}$$

It is easy to verify that $M = \tilde{M}$ and migration leaves the population distribution, $\underset{\sim}{c}$, invariant. The constant k $(0 \le k \le 1)$ represents the proportion of outbreeding. Observe that $k = 0$ corresponds to no migration, while $k = 1$ gives the Levene model, which includes no homing tendency. With two niches, both (6.71) and (6.73) have two independent parameters. The identification

$$m_1 = kc_2, \qquad m_2 = kc_1 \tag{6.74}$$

shows that in this case (6.73) is the most general model. For three or more niches, this is not so; the general backward migration matrix clearly has $N^2 - N = N(N-1)$ free parameters, while the random outbreeding and site homing model has only $1 + (N-1) = N$.

A condition equivalent to (6.72b) was derived by Maynard Smith (1970), whose h is $2(1-k)$. We owe the elegant form (6.72b) to Bulmer (1972).

Figure 6.2 shows that if *aa* is sufficiently deleterious relative to *Aa*, then *A* is protected. Explicitly, it suffices to have either $s_1 > m_1$ or $s_2 > m_2$. From (6.72b) and (6.74) we see that as the amount of dispersion, measured by k, decreases, the region of protected polymorphism in Figure 6.2 expands. As $k \to 0$, the hyperbolic boundary degenerates into the lines $s_2 = 0$, $s_1 \le 0$ and $s_1 = 0$, $s_2 \le 0$. Hence, provided $s_i > 0$ for some i, *A* is protected for sufficiently weak migration. This illustrates a general result proved in Section 6.2.

We proceed to discuss the case of recessive *A*. With $v_i = 1$ and $u_i = 1 - \sigma_i$, according to (6.72b), *a* is protected if

$$\frac{m_1}{\sigma_1} + \frac{m_2}{\sigma_2} < 1. \tag{6.75}$$

The normalized maximal left eigenvector of (6.71) reads

$$\underset{\sim}{U} = \begin{pmatrix} m_2/(m_1 + m_2) \\ m_1/(m_1 + m_2) \end{pmatrix}. \tag{6.76}$$

Therefore, the criterion (6.44) for protecting *A* becomes

$$m_2\sigma_1 + m_1\sigma_2 < 0. \tag{6.77}$$

We exhibit the region of protected polymorphism with fixed m_i in Figure 6.3.

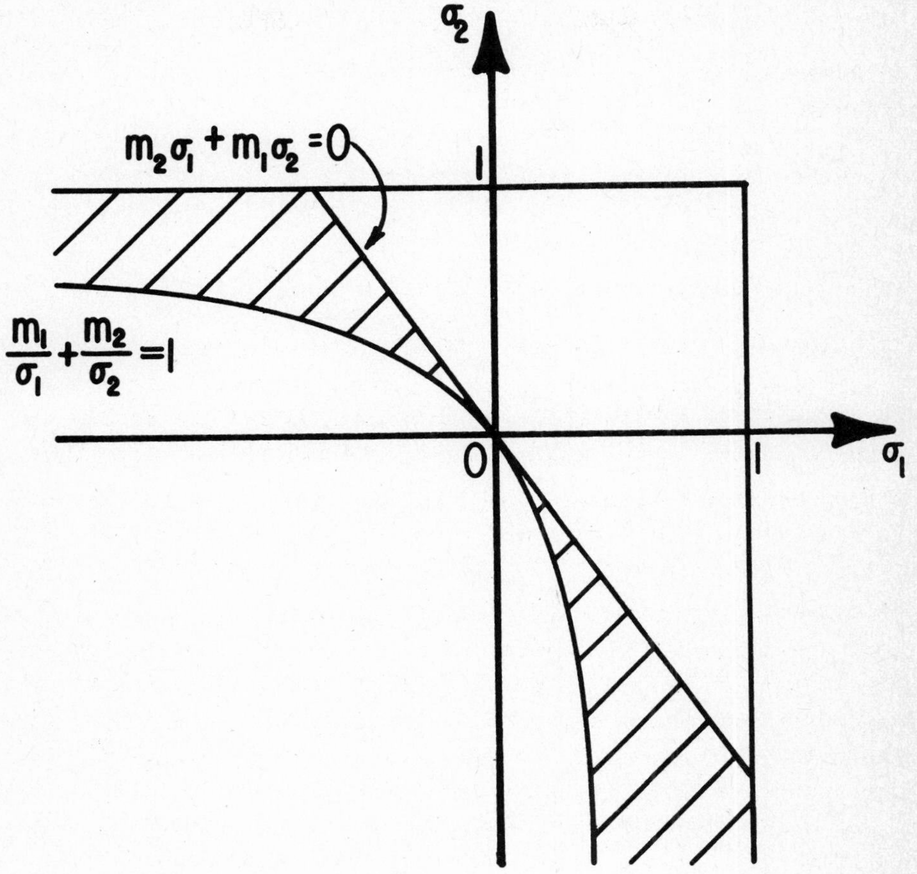

Fig. 6.3. The region of protected polymorphism with complete dominance (hatched).

Inserting (6.74) into (6.75) and (6.77) yields

$$k\left(\frac{c_2}{\sigma_1} + \frac{c_1}{\sigma_2}\right) < 1, \tag{6.78a}$$

$$c_1\sigma_1 + c_2\sigma_2 < 0. \tag{6.78b}$$

The condition (6.78b), which requires a negative mean selection coefficient, is due to Maynard Smith (1970). Again, as the amount of migration decreases, so does the stringency of (6.78).

Because of the possibility of selfing and the fact that it is pollen and seeds that disperse, most plant populations require modifications of our basic model. The interested reader may consult Nagylaki (1976b).

6.5 *Problems*

6.1. Derive the equilibrium structure exhibited in Figure 6.1.

6.2. Give an independent proof of global nonoscillating convergence for the island model without dominance.

6.3. To show that the matrices AB and BA have the same eigenvalues, note first that $\det(AB) = \det(BA)$. Therefore, zero is an eigenvalue of AB if and only if it is an eigenvalue of BA. Next, verify the identity

$$\lambda(\lambda I - BA)^{-1} = I + B(\lambda I - AB)^{-1}A,$$

where I is the identity matrix. Hence, for $\lambda \neq 0$ $(\lambda I - BA)^{-1}$ exists if and only if $(\lambda I - AB)^{-1}$ exists. So, $\det(\lambda I - BA) = 0$ if and only if $\det(\lambda I - AB) = 0$, proving the identity of the nonzero eigenvalues.

6.4. Demonstrate that migration does not change the allelic frequencies in the total (subdivided) population.

6.5. Deduce (6.59) and (6.60) from (6.54).

6.6. To establish that the arithmetic mean, $\bar{u}$, exceeds the harmonic mean, $\tilde{u}$, unless all the u_i are the same, use Schwarz's inequality to obtain

$$1 = \Big(\sum_i c_i\Big)^2 = \Big(\sum_i \sqrt{c_i/u_i}\,\sqrt{c_i u_i}\Big)^2 \leq \Big(\sum_i c_i/u_i\Big)\Big(\sum_i c_i u_i\Big) = \bar{u}/\tilde{u}.$$

Equality holds if and only if $\sqrt{c_i/u_i} = \beta\sqrt{c_i u_i}$, for some constant β. This requires that all the u_i be the same.

6.7. Derive (6.72).

7. X-LINKAGE

As discussed at the beginning of Section 3.2, X-linked loci are of particular interest and usefulness in human, mouse, and *Drosophila* genetics. We shall confine ourselves to discrete nonoverlapping generations; for the continuous time theory, the reader may refer to Nagylaki (1975c). In Section 7.1 we present the general formulation for selection and mutation with multiple alleles. We discuss the dynamics of selection with two alleles in Section 7.2, and treat multi-allelic mutation-selection balance in Section 7.3. The formulation of selection at a multiallelic autosomal locus in a dioecious population closely resembles the theory for X-linkage; we leave it for Problem 7.3.

7.1 *Formulation for Multiallelic Selection and Mutation*

We follow the notation and reasoning in Sections 3.2, 4.1, and 4.9. Let p_i and Q_{ij} represent the frequencies of A_i males and ordered A_iA_j females among zygotes. Thus, the frequency of A_i in female zygotes is

$$q_i = \sum_j Q_{ij}. \tag{7.1}$$

We designate by u_i and v_{ij} the probabilities that A_i males and A_iA_j females survive to reproductive age. The mean viability, v_i, of a female carrying A_i is given by

$$q_i v_i = \sum_j v_{ij} Q_{ij}, \tag{7.2}$$

and the average viabilities of males and females read

$$\overline{u} = \sum_i u_i p_i, \qquad \overline{v} = \sum_{ij} v_{ij} Q_{ij}. \tag{7.3}$$

Employing a tilde to indicate adult frequencies, we have

$$\tilde{p}_i = p_i u_i / \overline{u}, \tag{7.4a}$$

$$\tilde{Q}_{ij} = Q_{ij} v_{ij} / \overline{v}, \qquad \tilde{q}_i = q_i v_i / \overline{v}. \tag{7.4b}$$

We designate the frequency of the mating of an A_i male with an (ordered) A_jA_k female by $X_{i,jk}$. Hence,

$$\sum_{ijk} X_{i,jk} = 1. \tag{7.5}$$

Let $f_{i,jk}$ and $g_{i,jk}$ be the average number of males and females from an $A_i \times A_jA_k$ union. Although these fertilities will almost invariably be the same, we need not impose this restriction at this point. To incorporate mutation, suppose the probabilities that an A_i allele in a zygote appears as A_j in a gamete are R_{ij} and S_{ij} in males and females. Since there is some evidence that the mutation rates in man (Vogel, 1970) and mice (Searle, 1974) may be different in the two sexes, we shall not assume $R_{ij} = S_{ij}$. The genotypic proportions in zygotes in the next generation are

$$p_i' = \bar{f}^{-1} \sum_{klm} X_{k,lm} f_{k,lm} S_{li}, \tag{7.6a}$$

$$Q_{ij}' = \tfrac{1}{2}\bar{g}^{-1} \sum_{klm} X_{k,lm} g_{k,lm} (R_{ki}S_{lj} + R_{kj}S_{li}), \tag{7.6b}$$

where the mean fertilities for the birth of males and females read

$$\bar{f} = \sum_{klm} X_{k,lm} f_{k,lm}, \qquad \bar{g} = \sum_{klm} X_{k,lm} g_{k,lm}. \tag{7.7}$$

The identities

$$\sum_j R_{ij} = 1, \qquad \sum_j S_{ij} = 1, \tag{7.8}$$

immediately allow the verification of the normalizations in (7.6).

In general, (7.6b) cannot be further simplified. To elucidate the underlying biology of the model, we consider gene frequencies. From (7.6b) we find with the aid of (7.1) and (7.8)

$$q_i' = \tfrac{1}{2}\bar{g}^{-1} \sum_{klm} X_{k,lm} g_{k,lm} (R_{ki} + S_{li}). \tag{7.9}$$

The average numbers of females born to an A_i male, males born to an A_iA_j female, and females born to an A_iA_j female are given by

$$\tilde{p}_i a_i = \sum_{kl} X_{i,kl} g_{i,kl}, \tag{7.10a}$$

$$\tilde{Q}_{ij} b_{ij} = \sum_k X_{k,ij} f_{k,ij}, \tag{7.10b}$$

$$\tilde{Q}_{ij} c_{ij} = \sum_k X_{k,ij} g_{k,ij}. \tag{7.10c}$$

Therefore, the rates at which females carrying A_i give birth to males and females may be calculated from

$$\tilde{q}_i b_i = \sum_j \tilde{Q}_{ij} b_{ij} = \sum_{jk} X_{k,ij} f_{k,ij}, \qquad (7.11a)$$

$$\tilde{q}_i c_i = \sum_j \tilde{Q}_{ij} c_{ij} = \sum_{jk} X_{k,ij} g_{k,ij}. \qquad (7.11b)$$

From (7.7), (7.10a), and (7.11) we deduce at once that the mean fertilities are

$$\bar{a} = \sum_i \tilde{p}_i a_i = \bar{g} = \sum_i \tilde{q}_i c_i = \bar{c}, \qquad (7.12a)$$

$$\bar{b} = \sum_i \tilde{q}_i b_i = \bar{f}. \qquad (7.12b)$$

Since adults of the two sexes contribute equally to females in the next generation, the equality of the average rates at which males and females give birth to females, $\bar{a} = \bar{c}$, is expected. Substituting (7.10a), (7.11), and (7.12) into (7.6a) and (7.9), we obtain

$$p_i' = \bar{b}^{-1} \sum_k \tilde{q}_k b_k S_{ki}, \qquad (7.13a)$$

$$q_i' = \tfrac{1}{2}\bar{a}^{-1} \sum_k (\tilde{p}_k a_k R_{ki} + \tilde{q}_k c_k S_{ki}). \qquad (7.13b)$$

We define

$$x_i = u_i a_i, \qquad y_{ij} = v_{ij} b_{ij}, \qquad z_{ij} = v_{ij} c_{ij}. \qquad (7.14)$$

These quantities are the expected number of female progeny of a juvenile (i.e., pre-selection) A_i male, male progeny of a juvenile A_iA_j female, and female progeny of a juvenile A_iA_j female, and may be referred to as fitnesses. Observe that

$$\bar{x} = \sum_i x_i p_i = \sum_i u \tilde{p}_i a_i = \bar{u}\bar{a}, \qquad (7.15)$$

from (7.14), (7.4a), and (7.12a). The average contributions of females carrying A_i to males and females are given by

$$q_i y_i = \sum_j Q_{ij} y_{ij}, \qquad q_i z_i = \sum_j Q_{ij} z_{ij}. \qquad (7.16)$$

Using successively (7.16), (7.14), (7.4b), (7.11a), and (7.4b) again, we derive

$$q_i y_i = \sum_j \bar{v}\tilde{Q}_{ij} b_{ij} = \bar{v}\tilde{q}_i b_i = q_i v_i b_i. \tag{7.17}$$

Hence, and by identical reasoning on z_i, we get

$$y_i = v_i b_i, \qquad z_i = v_i c_i. \tag{7.18}$$

Utilizing (7.4b) and (7.12), the device employed in (7.15) yields

$$\bar{y} = \sum_i q_i y_i = \bar{v}\bar{b}, \tag{7.19a}$$

$$\bar{z} = \sum_i q_i z_i = \bar{v}\bar{c}. \tag{7.19b}$$

Although the factorization properties (7.15), (7.18), and (7.19) of the mean and allelic fitnesses are expected from the intuition concerning survival and reproduction, since they depend on the genotypic definitions (7.14), they require proof.

It remains only to substitute (7.4), (7.14), and (7.18) into (7.13) to derive our basic equations for the gene frequencies:

$$p_i' = q_i^{**}, \qquad q_i' = \tfrac{1}{2}(p_i^* + q_i^*), \tag{7.20}$$

where

$$p_i^* = \sum_k (p_k x_k / \bar{x}) R_{ki}, \tag{7.21a}$$

$$q_i^{**} = \sum_k (q_k y_k / \bar{y}) S_{ki}, \tag{7.21b}$$

$$q_i^* = \sum_k (q_k z_k / \bar{z}) S_{ki}. \tag{7.21c}$$

Note that the system (7.20), (7.21) is not complete: it depends on $X_{i,jk}$ and Q_{ij}. The interpretation of the starred allelic frequencies is immediate: q_i^{**}, q_i^*, and p_i^* are the frequencies of A_i in eggs producing males, eggs producing females, and sperm producing females. The analysis of random mating below will support this assertion. If, as is generally the case, the sex ratio at birth, λ, is independent of the parental genotypes, $f_{i,jk} = \lambda g_{i,jk}$, then $b_{ij} = \lambda c_{ij}$, so $y_{ij} = \lambda z_{ij}$, whence $q_i^* = q_i^{**}$.

To obtain a complete system of difference equations, let us suppose mating is random:

$$X_{i,jk} = \tilde{p}_i \tilde{Q}_{jk}. \tag{7.22}$$

Substituting (7.22) into (7.10) gives

$$a_i = \sum_{kl} \tilde{Q}_{kl} g_{i,kl}, \tag{7.23a}$$

$$b_{ij} = \sum_k \tilde{p}_k f_{k,ij}, \qquad c_{ij} = \sum_k \tilde{p}_k g_{k,ij}. \tag{7.23b}$$

Let us posit that the rates at which various unions produce females may be expressed as products of factors depending on the male and female genotypes:

$$g_{i,jk} = \beta_i \gamma_{jk}. \tag{7.24}$$

The corresponding restriction for the rates at which males are born is not required. Substituting (7.24) into (7.23) yields

$$a_i = \beta_i \bar{\gamma}, \qquad c_{ij} = \bar{\beta} \gamma_{ij}, \tag{7.25}$$

where

$$\bar{\beta} = \sum_i \beta_i \tilde{p}_i, \qquad \bar{\gamma} = \sum_{ij} \gamma_{ij} \tilde{Q}_{ij}. \tag{7.26}$$

Inserting (7.25) into (7.12a) and (7.11b) and comparing the result to (7.26), we find

$$\bar{a} = \bar{c} = \bar{\beta}\bar{\gamma}. \tag{7.27}$$

Equations (7.24), (7.25), and (7.27) lead to

$$g_{i,jk} = a_i c_{jk} / \bar{a}. \tag{7.28}$$

Substituting (7.22) and (7.28) into (7.6b) and recalling (7.4), (7.14), (7.15), (7.19b), (7.16), and (7.21), we deduce the generalized Hardy-Weinberg law,

$$Q'_{ij} = \tfrac{1}{2}(p_i^* q_j^* + p_j^* q_i^*). \tag{7.29}$$

The equations are conveniently expressed in terms of gametic frequencies. From (7.21a), (7.20), and (7.15),

$$p_i^{*\prime} = \sum_k (q_k^{**} x_k / \bar{x}') R_{ki}, \tag{7.30a}$$

with

$$\bar{x}' = \sum_i x_i q_i^{**}. \tag{7.30b}$$

Similarly, from (7.21b), (7.21c), (7.16), and (7.29) we derive

$$q_i^{**\prime} = \sum_k (q_k' y_k' / \bar{y}') S_{ki}, \tag{7.31a}$$

where

$$q_k' y_k' = \frac{1}{2} \sum_l (p_k^* q_l^* + p_l^* q_k^*) y_{kl}, \tag{7.31b}$$

$$q_i^{*\prime} = \sum_k (q_k' z_k' / \bar{z}') S_{ki}, \tag{7.32a}$$

where

$$q_k' z_k' = \frac{1}{2} \sum_l (p_k^* q_l^* + p_l^* q_k^*) z_{kl}, \tag{7.32b}$$

and, of course,

$$\bar{y}' = \sum_i q_i' y_i', \qquad \bar{z}' = \sum_i q_i' z_i'. \tag{7.33}$$

With the zygotic sex ratio independent of the parental genotypes, $q_i^* = q_i^{**}$, and (7.30) and (7.31) specify the evolution of the population.

Finally, if μ_{ij} and ν_{ij} are the mutation rates in males and females, with the convention that $\mu_{ii} = \nu_{ii} = 0$ for all i, we have

$$R_{ij} = \delta_{ij}(1 - \sum_k \mu_{ik}) + \mu_{ij}, \tag{7.34a}$$

$$S_{ij} = \delta_{ij}(1 - \sum_k \nu_{ik}) + \nu_{ij}. \tag{7.34b}$$

Therefore, (7.30a) may be rewritten as

$$p_i^{*\prime} = (q_i^{**} x_i / \bar{x}')(1 - \sum_k \mu_{ik}) + \sum_k (q_k^{**} x_k / \bar{x}') \mu_{ki}, \tag{7.35}$$

with similar equations for q_i^* and q_i^{**}.

In the absence of selection, (7.20) and (7.21) reduce to the linear system

$$p_i' = \sum_k q_k S_{ki}, \tag{7.36a}$$

$$q_i' = \frac{1}{2} \sum_k (p_k R_{ki} + q_k S_{ki}). \tag{7.36b}$$

Recalling (7.34), it is easy to solve the diallelic case explicitly for $p_1(t)$ and $q_1(t)$. If $\mu_{ij}, \nu_{ij} \ll 1$, one finds at equilibrium

$$\hat{p}_1 \approx \hat{q}_1 \approx (\mu_{21} + 2\nu_{21})/(\tilde{\mu} + 2\tilde{\nu}), \tag{7.37a}$$

where

$$\tilde{\mu} = \mu_{12} + \mu_{21}, \qquad \tilde{\nu} = \nu_{12} + \nu_{21}. \tag{7.37b}$$

For weak mutation, one expects the male and female gene frequencies at equilibrium to be equal. The factor 2 appears in (7.37a) because females have two X-chromosomes to the males' one. The ultimate rate of approach to the equilibrium is

$$[1 - \tfrac{1}{3}(\tilde{\mu} + 2\tilde{\nu})]^t \approx e^{-(\tilde{\mu} + 2\tilde{\nu})t/3}, \tag{7.38}$$

which displays the same factor 2. For $\mu_{ij} = \nu_{ij}$, (7.37) and (7.38) reduce to the autosomal case, (2.22) and (2.23).

If there is no mutation, and the juvenile sex ratio is genotype-independent, (7.30) and (7.31a) simplify to

$$p_i^{*\prime} = q_i^* x_i \,/\, \bar{x}', \qquad q_i^{*\prime} = q_i' y_i' \,/\, \bar{y}', \tag{7.39}$$

in which

$$\bar{x}' = \sum_i x_i q_i^*, \tag{7.40}$$

and $q_i' y_i'$ and $\bar{y}'$ are given by (7.31b) and (7.33). Sprott (1957) has discussed the stability of multiallelic polymorphisms of this system, and Cannings (1968a) showed that these polymorphisms, if they exist, may be calculated by matrix inversion, as for the autosomal case.

7.2 *Selection with Two Alleles*

With two alleles, no mutation, and genotype-independent zygotic sex ratio, (7.31b), (7.33), (7.39), and (7.40) yield the recursion relations for the gene frequency ratios (Haldane, 1926)

$$\xi = p_1^* / p_2^*, \qquad \eta = q_1^* / q_2^*: \tag{7.41}$$

$$\xi' = \rho\eta, \qquad \eta' = (2\sigma\xi\eta + \xi + \eta)/(\xi + \eta + 2\tau), \tag{7.42}$$

where the fitnesses appear only in the ratios

$$\rho = x_1/x_2, \qquad \sigma = y_{11}/y_{12}, \qquad \tau = y_{22}/y_{12}. \tag{7.43}$$

We shall suppose that ρ, σ, and τ are constant.

If one of the alleles is lethal in males, it is easy to solve (7.42) for $\xi(t)$ and $\eta(t)$ (Problem 7.2). Haldane (1924) obtained the special case of this solution for recessives lethal also in females.

The mapping (7.42) of the first quadrant of the $\xi\eta$-plane into itself has the trivial fixed points (0,0) and (∞,∞), corresponding to the fixation of a and A, respectively. If it exists (and a degenerate situation treated below is excluded), the unique polymorphic equilibrium

$$\hat{\xi} = \rho\hat{\eta}, \qquad \hat{\eta} = (\rho + 1 - 2\tau)/(\rho + 1 - 2\rho\sigma) \tag{7.44}$$

can be located directly from (7.42). Manifestly, $0 < \hat{\eta} < \infty$, whence $0 < \hat{\xi} < \infty$, if and only if either

$$\tau < \tfrac{1}{2}(1+\rho) \quad \text{and} \quad \sigma < \tfrac{1}{2}(1+\rho^{-1}), \tag{7.45a}$$

or

$$\tau > \tfrac{1}{2}(1+\rho) \quad \text{and} \quad \sigma > \tfrac{1}{2}(1+\rho^{-1}). \tag{7.45b}$$

Thus, (7.45) is necessary and sufficient for the existence of a nontrivial equilibrium.

Now, $\sigma,\tau << 1$ corresponds to extreme heterozygote advantage, which we expect to lead to a stable equilibrium. Therefore, intuition suggests that, as Bennett (1957) has proved, (7.44) is locally stable whenever (7.45a) holds. The analogous underdominance argument disposes us to guess that (7.45b) implies instability of (7.44).

Let us show that (7.45a) guarantees a protected polymorphism (Kimura, 1960; see Crow and Kimura, 1970, pp. 278-281). Near the origin, (7.42) has the linearization

$$\underset{\sim}{Y}' = B\underset{\sim}{Y},$$

where

$$\underset{\sim}{Y} = \begin{pmatrix} \xi \\ \eta \end{pmatrix}, \qquad B = \begin{pmatrix} 0 & \rho \\ 1/(2\tau) & 1/(2\tau) \end{pmatrix}.$$

From (6.32) we find immediately that the maximal eigenvalue of the irreducible nonnegative matrix B exceeds unity if

$$\tau < \tfrac{1}{2}(1+\rho). \tag{7.46}$$

By the argument in Section 6.2, (7.46) suffices for protection of the allele A. We can find the condition for maintaining a by the interchange $A \leftrightarrow a$. According to (7.43), this corresponds to $\rho \leftrightarrow \rho^{-1}$ and $\sigma \leftrightarrow \tau$. Applying these transformations to (7.46) produces the second half of (7.45a).

We describe below the global behavior of (7.42). Case 4 and part of Case 5 were first analyzed by Cannings (1967) and the remainder by Palm (1974). The reader may consult their papers for proofs. For convenience we set

$$\alpha = \sigma, \qquad \beta = \tau\rho^{-1}, \qquad \gamma = \tfrac{1}{2}(1+\rho^{-1}). \tag{7.47}$$

1. $\alpha \leq \gamma \leq \beta$, $\alpha \neq \beta$:

   Since (7.45) is violated, only the trivial equilibria exist; *A* is lost.

2. $\alpha \geq \gamma \geq \beta$, $\alpha \neq \beta$:

   Again, (7.45) fails, so a polymorphism does not exist, but now *A* is fixed.

3. $\alpha = \beta = \gamma$:

   In this degenerate case we can easily check that all points on the line $(\rho\eta, \eta)$ through the origin are equilibria; there is global convergence to this line.

4. $\alpha, \beta < \gamma$:

   Now (7.45a) is satisfied, and there is global convergence to the polymorphism (7.44).

5. $\alpha, \beta > \gamma$:

   Owing to (7.45b), the nontrivial equilibrium (7.44) is unstable. There is convergence to (7.44) from its unidimensional stable manifold, *S*. If the population is initially not on *S*, one of the alleles will be ultimately lost. (As usual, this will always happen in practice.) The curve *S* through (7.44) separates the regions of attraction of the trivial equilibria. If $\eta(0), \eta(1) < \hat{\eta}$, then *A* is lost; if $\eta(0), \eta(1) > \hat{\eta}$, *A* is fixed.

The following special cases are illuminating.

(a) No selection in males:

   With $\rho = 1$, the male and female gene frequencies are ultimately equal. In the five cases above, we have $\alpha = \sigma$, $\beta = \tau$, and $\gamma = 1$.

The fate of the population (though not its evolution) is the same as if the locus were autosomal and all individuals had the female fitnesses.

(b) No selection in females:

Clearly, $\alpha = 1$, $\beta = \rho^{-1}$, and $\gamma = \frac{1}{2}(1+\rho^{-1})$. No polymorphism is possible, the eventual fate of the population being that of a haploid one with the male fitnesses.

(c) A dominant:

Using $\sigma = 1$ in (7.45a) yields $\rho < 1$ and $\tau < 1$. Hence (excluding the case without selection), for a stable polymorphism the recessive allele must be advantageous in males and deleterious in females.

(d) No dominance:

From (7.43) we get $\sigma + \tau = 2$, whence the necessary and sufficient condition (7.45a) for a stable polymorphism in the presence of selection becomes

$$\frac{1}{2}(3-\rho) < \sigma < \frac{1}{2}(1+\rho^{-1}).$$

This can be satisfied only if A is favored in males and harmful in females ($\rho > 1$ and $\sigma < 1$) or *vice versa* ($\rho < 1$ and $\sigma > 1$).

(e) No multiplicative dominance:

Now $\sigma\tau = 1$, so (7.45a) simplifies to

$$2(1+\rho)^{-1} < \sigma < \frac{1}{2}(1+\rho^{-1}),$$

from which the conclusion (d) follows.

7.3 *Mutation-Selection Balance*

We shall relate at equilibrium the frequency of mutant alleles in males and females to the mean selection coefficients and mutation rates in the two sexes, as we did for an autosomal locus in Section 4.9. We suppose that the juvenile sex ratio is independent of the parental genotypes, and choose A_1 as the normal allele. We make no assumption about fertilities or the mating system. From (7.20), (7.21), and (7.34) we obtain at equilibrium

$$p_1 = (q_1 y_1 / \bar{y})(1-\nu) + \sum_i (q_i y_i / \bar{y})\nu_{i1}, \tag{7.48a}$$

$$q_1 = \tfrac{1}{2}[(p_1 x_1 / \bar{x})(1-\mu) + (q_1 y_1 / \bar{y})(1-\nu) + \sum_i (p_i x_i / \bar{x})\mu_{i1} + \sum_i (q_i y_i / \bar{y})\nu_{i1}], \tag{7.48b}$$

where

$$\mu = \sum_i \mu_{1i}, \qquad \nu = \sum_i \nu_{1i} \tag{7.49}$$

are the total forward mutation rates for A_1 in males and females.

We consider only deleterious mutants, so we choose the fitnesses $x_i = 1 - r_i$, $y_{ij} = 1 - s_{ij}$, with $r_i \geq 0$, $s_{ij} \geq 0$, $r_1 = s_{11} = 0$, and the biologically reasonable restriction that r_i and s_{1i} cannot both vanish for $i > 1$. For recessives, $s_{1i} = 0$, but it would be highly unusual to have $r_i = 0$ for $i > 1$. For weak mutation, which we may define by $\mu_{1i} + 2\nu_{1i} \ll r_i + 2s_{1i}$ for $i > 1$, the autosomal analysis of Section 4.9 leads us to expect that mutant frequencies will be ratios of mutation rates to selection coefficients. Then the reverse mutation terms, the sums in (7.48), may be neglected to the first order in mutation rates.

To proceed, we calculate the fitnesses in the dominant terms of (7.48). We have

$$\bar{x} = 1 - \sum_i r_i p_i = 1 - (1-p_1)\bar{r}, \tag{7.50a}$$

where

$$\bar{r} = (1-p_1)^{-1} \sum_i r_i p_i \tag{7.50b}$$

is the mean selection coefficient of the mutants in males. For the females, notice that for $i > 1$

$$q_i = Q_{i1} + \sum_{j>1} Q_{ij} \approx Q_{i1}, \tag{7.51}$$

since $Q_{ij} \ll Q_{i1}$ unless the inbreeding coefficient is quite substantial. Therefore,

$$q_1 y_1 = \sum_i (1-s_{1i})Q_{1i} \approx q_1 - \sum_i s_{1i} q_i = q_1 - (1-q_1)\bar{s}, \tag{7.52a}$$

where

$$\bar{s} = (1-q_1)^{-1} \sum_i q_i s_{1i} \tag{7.52b}$$

is the mean selection coefficient of normal-mutant females. Also,

$$\bar{y} = \sum_{ij} (1 - s_{ij}\, Q_{ij} \approx 1 - 2\sum_{i} s_{1i}Q_{1i} \approx 1 - 2(1 - q_1)\bar{s}. \qquad (7.53)$$

Substituting (7.50a), (7.52a), and (7.53) into (7.48) yields

$$p_1 \approx \left[\frac{q_1 - (1 - q_1)\bar{s}}{1 - 2(1 - q_1)\bar{s}}\right](1 - \nu), \qquad (7.54a)$$

$$q_1 \approx \frac{1}{2}\left[\frac{p_1}{1 - (1 - p_1)\bar{r}}\right](1 - \mu) + \frac{1}{2}\left[\frac{q_1 - (1 - q_1)\bar{s}}{1 - 2(1 - q_1)\bar{s}}\right](1 - \nu). \qquad (7.54b)$$

Recalling that $1 - p_1$ and $1 - q_1$ are of the order of mutation rates, to first order (7.54) becomes

$$p_1 \approx q_1 + (1 - q_1)\bar{s} - \nu, \qquad (7.55a)$$

$$2q_1 \approx p_1 + q_1 + (1 - p_1)\bar{r} + (1 - q_1)\bar{s} - \mu - \nu. \qquad (7.55b)$$

Inserting (7.55a) into (7.55b), we find immediately

$$1 - q_1 \approx \frac{\mu + (2 - \bar{r})\nu}{\bar{r} + 2\bar{s} - \bar{r}\bar{s}}, \qquad (7.56a)$$

and substituting (7.56a) into (7.55a) gives

$$1 - p_1 \approx \frac{(1 - \bar{s})\mu + 2\nu}{\bar{r} + 2\bar{s} - \bar{r}\bar{s}}. \qquad (7.56b)$$

Several special cases of our general result, (7.56), are of interest.

If the total mutation rate is the same in the two sexes, (7.56) reduces to

$$1 - p_1 \approx \frac{(3 - \bar{s})\mu}{\bar{r} + 2\bar{s} - \bar{r}\bar{s}}, \qquad 1 - q_1 \approx \frac{(3 - \bar{r})\mu}{\bar{r} + 2\bar{s} - \bar{r}\bar{s}}. \qquad (7.57)$$

With equal total mutation rates and mean selection coefficients in the two sexes, (7.57) simplifies to

$$1 - p_1 \approx 1 - q_1 \approx \mu / \bar{r}. \qquad (7.58)$$

If A_1 is dominant, $\bar{s} = 0$, so (7.56) reads

$$1 - p_1 \approx (\mu + 2\nu)/\bar{r}, \qquad 1 - q_1 \approx [\mu + (2 - \bar{r})\nu]/\bar{r}. \qquad (7.59)$$

Haldane (1935) derived the special case of (7.59) with two alleles and generalized Hardy-Weinberg proportions, (7.29). His formulae are for p_2^* and q_2^* with the definitions of μ and ν interchanged. If the mutants are lethal, (7.59) becomes

$$1-p_1 \approx \mu+2\nu, \qquad 1-q_1 \approx \mu+\nu.$$

Finally, for weak selection, (7.56) becomes

$$1-p_1 \approx 1-q_1 \approx \frac{\mu+2\nu}{\bar{r}+2\bar{s}},$$

both the approximate equality of p_1 and q_1 and the factors of 2 being expected.

7.4 *Problems*

7.1. Verify (7.37) and (7.38).

7.2. Solve (7.42) for $\rho = 0$ (A lethal in males). Show that A is lost for $\tau \geq 1/2$, but there is a globally stable polymorphism for $\tau < 1/2$.

7.3. Give a complete formulation of selection in a dioecious population with discrete nonoverlapping generations. Consider a single autosomal locus with multiple alleles. In terms of mating frequencies, fertilities, and viabilities, write the most general recursion relation for the zygotic genotypic frequencies. Show that with suitably defined fitnesses the zygotic allelic frequencies always satisfy

$$p_i' = \frac{1}{2}\left(\frac{p_i w_i}{\bar{w}} + \frac{q_i y_i}{\bar{y}}\right),$$

$$q_i' = \frac{1}{2}\left(\frac{p_i x_i}{\bar{x}} + \frac{q_i z_i}{\bar{z}}\right).$$

Reduce the equations for the zygotic proportions with the additional assumptions of random mating and multiplicative fertilities. Prove that the zygotic frequencies are in generalized Hardy-Weinberg proportions in terms of the frequencies p_i^*, p_i^{**}, q_i^*, q_i^{**} of A_i in sperm producing males, sperm producing females, eggs producing males, eggs producing females. Demonstrate that with the sex ratio independent of parental genotypes $p_i^{**} = p_i^*$, $q_i^{**} = q_i^*$, and the zygotic frequencies are the same in the two sexes after one generation of panmixia. Show that in this simple standard case the model is completely specified by the equations

$$p_i^{*\prime} = B_i / B, \qquad q_i^{*\prime} = C_i / C,$$

$$B_i = \frac{1}{2} \sum_j (p_i^* q_j^* + p_j^* q_i^*) w_{ij}, \qquad B = \sum_i B_i,$$

$$C_i = \frac{1}{2} \sum_j (p_i^* q_j^* + p_j^* q_i^*) y_{ij}, \qquad C = \sum_i C_i.$$

If there is no selection in females, prove that these equations reduce at equilibrium to those for an autosome with the male fitnesses.

For some analyses and references, see Bodmer (1965).

8. TWO LOCI

We are interested in the two-locus system because some characters are determined by two loci and the understanding of this system is essential for the analysis of multilocus traits. The latter, exhibiting continuous or almost continuous variation, are of the greatest evolutionary significance. In view of our second motivation, it will be useful to devote Section 8.1 to establishing generalized Hardy-Weinberg proportions and deriving the recursion relations satisfied by the gametic frequencies for an arbitrary number of autosomal loci. We shall specialize our equations to two multiallelic loci and analyze this system in Section 8.2. Examples concerning two diallelic loci presented in Section 8.3 will further elucidate two-locus behavior. Section 8.4 deals with two multiallelic loci in continuous time.

8.1 *General Formulation for Multiple Loci*

We consider an arbitrary number of autosomal loci, and posit random mating and multiplicative fertilities. In the process of demonstrating that panmixia implies random union of gametes, we shall deduce the difference equations for the gametic frequencies. Our treatment slightly generalizes that of Bodmer (1965).

We designate the frequencies of the gamete G_i in the gametic output of males and females of generation t by $p_i(t)$ and $q_i(t)$. Let the genotype G_iG_j have frequencies $P_{ij}(t)$ and $Q_{ij}(t)$ in adult males and females in generation t. We assume that the fertility of the union of a male of genotype G_iG_j with a female of genotype G_kG_l has the form $\alpha_{ij}\beta_{kl}$ for some α_{ij} and β_{kl}. In the rare event that parental genotypes should influence the sex ratio, we would have to distinguish gametic frequencies in sperm and eggs and fertilities for producing males and females. It would be easy fo carry out this extension in the manner of Section 7.1 and Problem 7.3. Let $R_{i,jk}$ and $S_{i,jk}$ denote the probabilities that males and females of genotype G_jG_k produce a gamete G_i.

Since

$$\sum_i R_{i,jk} = 1 \quad \text{and} \quad \sum_i S_{i,jk} = 1,$$

the frequency of the genotype G_iG_j in zygotes in generation $t+1$ reads

$$T_{ij}(t+1) = (\bar{\alpha}\bar{\beta})^{-1} \sum_{klmn} P_{kl}(t)Q_{mn}(t)\alpha_{kl}\beta_{mn}\frac{1}{2}(R_{i,kl}S_{j,mn} + R_{j,kl}S_{i,mn}),$$

where

$$\bar{\alpha} = \sum_{ij} P_{ij}\alpha_{ij}, \qquad \bar{\beta} = \sum_{ij} Q_{ij}\beta_{ij}.$$

This has the generalized Hardy-Weinberg form

$$T_{ij}(t+1) = \frac{1}{2}[\, p_i(t)q_j(t) + p_j(t)q_i(t)\,] \tag{8.1}$$

in terms of the gametic frequencies

$$p_i = \bar{\alpha}^{-1} \sum_{jk} P_{jk}\alpha_{jk}R_{i,jk}, \qquad q_i = \bar{\beta}^{-1} \sum_{jk} Q_{jk}\beta_{jk}S_{i,jk}. \tag{8.2}$$

If the male and female viabilities are u_{ij} and v_{ij}, the adult frequencies in generation $t+1$ will be

$$P_{ij} = u_{ij}T_{ij} / \bar{u}, \qquad Q_{ij} = v_{ij}T_{ij} / \bar{v}, \tag{8.3}$$

with the mean viabilities

$$\bar{u} = \sum_{ij} T_{ij}u_{ij}, \qquad \bar{v} = \sum_{ij} T_{ij}v_{ij}. \tag{8.4}$$

Writing (8.2) at time $t+1$ and substituting (8.3) yields

$$p_i(t+1) = (\bar{\alpha}\bar{u})^{-1} \sum_{jk} \alpha_{jk}u_{jk}T_{jk}(t+1)R_{i,jk}, \tag{8.5a}$$

$$q_i(t+1) = (\bar{\beta}\bar{v})^{-1} \sum_{jk} \beta_{jk}v_{jk}T_{jk}(t+1)S_{i,jk}. \tag{8.5b}$$

By virtue of (8.1), the system (8.5) is a set of recursion relations for the gametic frequencies. Since the latter are normalized, from (8.5) we obtain

$$\bar{\alpha}\bar{u} = \overline{\alpha u} = \sum_{jk} \alpha_{jk}u_{jk}T_{jk}, \tag{8.6a}$$

$$\bar{\beta}\bar{v} = \overline{\beta v} = \sum_{jk} \beta_{jk}v_{jk}T_{jk}. \tag{8.6b}$$

Relations equivalent to (8.6) were proved directly in special cases in Sections 4.1 and 7.1. The products $\alpha_{ij}u_{ij}$ and $\beta_{ij}v_{ij}$ represent the male and female fitnesses. If these are proportional, $\alpha_{ij}u_{ij}$ =

(const)$v_{ij}\beta_{ij}$, and the linkage maps in the two sexes are the same, $R_{i,jk} = S_{i,jk}$, then (8.5) shows that after one generation of panmixia the gametic frequencies in the two sexes will be the same, and (8.5) reduces after two generations to

$$p_i' = \bar{w}^{-1} \sum_{jk} w_{jk} R_{i,jk} p_j p_k, \tag{8.7a}$$

where $w_{ij} = \alpha_{ij} u_{ij}$ and

$$\bar{w} = \sum_{jk} w_{jk} p_j p_k. \tag{8.7b}$$

Henceforth, we shall confine ourselves to the monoecious model (8.7). Observe that for a single locus the Mendelian formula

$$R_{i,jk} = \tfrac{1}{2}(\delta_{ij} + \delta_{ik})$$

immediately simplifies (8.7) to (4.25).

8.2 *Analysis for Two Multiallelic Loci*

As in Section 3.3, we denote the alleles at the two loci by A_i and B_j, and designate the frequency of the gamete A_iB_j in the gametic output of generation t by $P_{ij}(t)$. Let $w_{ij,kl}$ $(= w_{kl,ij})$ represent the fitness of the genotype A_iB_j / A_kB_l. The fitnesses of A_iB_j / A_kB_l and A_iB_l / A_kB_j are almost invariably the same, i.e., $w_{ij,kl} = w_{il,kj}$. This means that there is no *position effect*: given the composition of the genotype, the arrangement of the alleles is irrelevant. Nevertheless, we shall not impose this restriction on the fitnesses. We denote the probability of drawing an A_iB_j gamete from an A_kB_l / A_mB_n individual by $R_{ij;kl,mn}$. In our new notation, (8.7) reads

$$P_{ij}' = \bar{w}^{-1} \sum_{klmn} w_{kl,mn} P_{kl} P_{mn} R_{ij;kl,mn}, \tag{8.8a}$$

with

$$\bar{w} = \sum_{klmn} w_{kl,mn} P_{kl} P_{mn}. \tag{8.8b}$$

Suppose c is the recombination fraction between the A and B loci. To calculate R, first consider drawing A_i, and then take into account recombination:

$$R_{ij;kl,mn} = \tfrac{1}{2}\delta_{ik}[(1-c)\delta_{jl} + c\delta_{jn}] + \tfrac{1}{2}\delta_{im}[(1-c)\delta_{jn} + c\delta_{jl}]. \tag{8.9}$$

Since the sum in (8.8a) is symmetric under the simultaneous interchanges $k \leftrightarrow m$ and $l \leftrightarrow n$, the two terms in (8.9) contribute equally. Therefore, substituting (8.9) into (8.8a) leads to our fundamental recursion relation,

$$\bar{w}P'_{ij} = P_{ij}w_{ij} - cD_{ij}, \tag{8.10}$$

where the linkage disequilibria, D_{ij}, are defined by

$$D_{ij} = \sum_{kl} (w_{ij,kl}P_{ij}P_{kl} - w_{il,kj}P_{il}P_{kj}), \tag{8.11}$$

and

$$w_{ij} = \sum_{kl} w_{ij,kl}P_{kl} \tag{8.12}$$

is the fitness of the gamete A_iB_j.

Kimura (1956a) deduced the differential equations for the gametic frequencies for two diallelic loci in a continuous-time Hardy-Weinberg model. Lewontin and Kojima (1960) obtained the corresponding difference equations for discrete nonoverlapping generations. For a more concrete derivation of (8.10), consult Kimura and Ohta (1971).

From (8.10) we find that the change in the frequency of A_iB_j is given by

$$\bar{w}\Delta P_{ij} = P_{ij}(w_{ij} - \bar{w}) - cD_{ij}. \tag{8.13}$$

We infer at once from (8.11) that the linkage disequilibria satisfy

$$\sum_i D_{ij} = 0, \qquad \sum_j D_{ij} = 0. \tag{8.14}$$

Note that in each term of the sum in (8.11) the two genotypes are related to each other by crossing over. As in Section 3.3, we define linkage equilibrium by the random combination of alleles within gametes:

$$P_{ij} = p_iq_j, \tag{8.15}$$

for all i and j, where

$$p_i = \sum_j P_{ij} \qquad \text{and} \qquad q_j = \sum_i P_{ij} \tag{8.16}$$

are the frequencies of A_i and B_j in gametes. Substituting (8.15) into (8.11), we see readily that without position effect, in linkage equilibrium, $D_{ij} = 0$ for all i and j. In the absence of selection, we may choose all the fitnesses to be unity, in which case (8.11) immediately reduces to (3.28), and (8.13) simplifies to (3.31).

We shall now analyze the dynamics of the two-locus system for weak selection (Nagylaki, 1976). The concepts are the same as those following (4.150) for a single locus with overlapping generations. Random combination now refers to linkage equilibrium, rather than Hardy-Weinberg proportions, and our measures of departure are the D_{ij}, rather than the Q_{ij} of (4.149). Our treatment and its interpretation follow Section 4.10.

As in (4.58), we may take

$$w_{ij,kl} = 1 + O(s) \tag{8.17}$$

as the selection intensity $s \to 0$. Our slow-selection analysis will apply if selection is much weaker than recombination, $s \ll c$. Since most pairs of loci are either on different chromosomes ($c = 1/2$) or quite far apart on the same chromosome (say, $c \gg 0.01$), and, excluding lethals, selection coefficients rarely exceed several percent, we expect our treatment to possess wide applicability. At this point, the fitnesses may depend on the gametic frequencies and time in an arbitrary manner.

To show that the linkage disequilibria, D_{ij}, settle down rapidly to approximate constancy, we must first derive a difference equation for them. From (8.10) we obtain with the aid of (8.17)

$$P'_{ij} = P_{ij} - cD_{ij} + O(s). \tag{8.18}$$

Rewriting (8.11) for D'_{ij}, substituting (8.17) and (8.18), and employing (8.14), we find

$$D'_{ij} = (1-c)D_{ij} + O(s).$$

Hence,

$$D'_{ij} = (1-c)D_{ij} + sg_{ij}(\underset{\sim}{P},t), \tag{8.19}$$

where g_{ij} is a complicated function of fitnesses (and therefore possibly of time) and gametic frequencies, which is uniformly bounded as $s \to 0$. Iterating (8.19) yields

$$D_{ij}(t) = D_{ij}(0)(1-c)^t + s(1-c)^t \sum_{\tau=1}^{t} (1-c)^{-\tau} g_{ij}[\underset{\sim}{P}(\tau-1),\tau-1]. \tag{8.20}$$

(The sum is absent for $t = 0$.)

We define t_1 as the shortest time such that

$$|D_{ij}(0)|(1-c)^{t_1} \le s \tag{8.21}$$

for all i, j. If the population starts sufficiently close to linkage equilibrium, then $t_1 = 0$. Otherwise, (8.21) gives the conservative estimate $t_1 \approx \ln s / \ln(1-c)$. For tight linkage ($s << c << 1$), this reduces to $t_1 \approx -c^{-1} \ln s$. With loose linkage or independent assortment, t_1 will be usually no more than 5 or 10 generations. From (8.20) we conclude that the linkage disequilibria are reduced to the order of the selection intensity in this short time: $D_{ij}(t) = O(s)$, $t \geq t_1$. Thus, we have $D_{ij}(t) = sD_{ij}^0(t)$, with $D_{ij}^0(t) = O(1)$, $t \geq t_1$. Summing (8.10) over i and j and using (8.14) produces the usual recursion relations for the gene frequencies, so during the time t_1 the gene frequency change is very small, roughly $st_1 \approx s \ln s / \ln(1-c)$. For $t \geq t_1$, (8.19) can be reexpressed as

$$\Delta D_{ij}^0 = -cD_{ij}^0 + g_{ij}(\underset{\sim}{P}, t). \tag{8.22}$$

From (8.18) we infer that

$$\Delta P_{ij} = O(s), \qquad t \geq t_1. \tag{8.23}$$

Equation (8.22) informs us that

$$\Delta[D_{ij}^0 - c^{-1} g_{ij}(\underset{\sim}{P}, t)] = -cD_{ij}^0 + g_{ij}(\underset{\sim}{P}, t) - c^{-1} \Delta g_{ij}(\underset{\sim}{P}, t). \tag{8.24}$$

We decompose the change in g_{ij} into parts due to its dependence on the gametic frequencies and on time:

$$\Delta g_{ij}(\underset{\sim}{P}, t) = \{g_{ij}[\underset{\sim}{P}(t+1), t+1] - g_{ij}[\underset{\sim}{P}(t), t+1]\} + \{g_{ij}[\underset{\sim}{P}(t), t+1] - g_{ij}[\underset{\sim}{P}(t), t]\}. \tag{8.25}$$

In view of (8.23), the first brace is of $O(s)$. Since the selection term in (8.19) is sg_{ij}, if we assume that the explicit time dependence of the fitnesses is of $O(s^2)$, the second brace in (8.25) will also be of $O(s)$. Then $\Delta g_{ij} = O(s)$, and (8.24) becomes

$$\Delta[D_{ij}^0 - c^{-1} g_{ij}(\underset{\sim}{P}, t)] = -c[D_{ij}^0 - c^{-1} g_{ij}(\underset{\sim}{P}, t)] + O(s). \tag{8.26}$$

Comparing (8.19) and (8.26), we choose t_2 ($\geq t_1$) as the shortest time such that

$$|D_{ij}^0(t_1) - c^{-1} g_{ij}[\underset{\sim}{P}(t_1), t_1]|(1-c)^{t_2 - t_1} \leq s. \tag{8.27}$$

From (8.27) and an equation like (8.20) for the bracket in (8.26) we obtain

$$D_{ij}^0(t) = c^{-1} g_{ij}(\underset{\sim}{P}, t) + O(s), \qquad t \geq t_2.$$

Now $\Delta g_{ij} = O(s)$ implies $\Delta D_{ij}^0 = O(s)$, whence

$$\Delta D_{ij}(t) = O(s^2), \qquad t \geq t_2. \tag{8.28}$$

Crudely, (8.27) tells us that $t_2 - t_1 \approx t_1$, so $t_2 \approx 2t_1$. Thus, in a short time, generally less than 10 or 20 generations, the linkage disequilibria become nearly constant. This important qualitative observation was first made by Kimura (1965) in a model with two alleles at each locus. Conley (1972) was the first to prove the analogue of (8.28) in a diallelic, continuous-time, Hardy-Weinberg model. Since the amount of gene frequency change during the second period, $t_1 \leq t < t_2$, is approximately the same as during the first, at time t_2 the population is still very far from gene frequency equilibrium. The allelic frequencies require a time $t_3 \approx s^{-1}$ to approach equilibrium. The change in the linkage disequilibria during the third period, however, is quite small, viz., $s^2(t_3 - t_2) \approx s^2 t_3 \approx s$.

Before deducing the implications of (8.28) for the change in mean fitness, we shall show that, after a time t_1, the population evolves approximately as if it were in linkage equilibrium, the difference between the exact gametic frequencies and those of the much simpler hypothetical system on the linkage equilibrium surface being of $O(s)$. At time t_1, the exact allelic frequencies are $p_i(t_1)$ and $q_i(t_1)$, and these evolve according to the complicated laws $p_i(t)$ and $q_i(t)$, calculated from (8.10). On the linkage equilibrium surface, there is a unique point with allelic frequencies $\pi_i(t_1) = p_i(t_1)$ and $\rho_i(t_1) = q_i(t_1)$, which evolve according to the much simpler laws $\pi_i(t)$ and $\rho_i(t)$, calculated by imposing linkage equilibrium on (8.10). We wish to prove that $P_{ij}(t) = \pi_i(t)\rho_j(t) + O(s)$, $t \geq t_1$.

Substituting (8.17) into (8.11) gives

$$D_{ij} = P_{ij} - p_i q_j + O(s). \tag{8.29}$$

Since $D_{ij}(t) = O(s)$ for $t \geq t_1$, (8.29) informs us that

$$P_{ij}(t) = p_i(t)q_j(t) + O(s), \qquad t \geq t_1. \tag{8.30}$$

By virtue of (8.17), we write

$$w_{ij,kl}(\underset{\sim}{P},t) = 1 + s u_{ij,kl}(\underset{\sim}{P},t), \tag{8.31a}$$

whence

$$w_{ij}(\underset{\sim}{P},t) = 1 + s u_{ij}(\underset{\sim}{P},t), \tag{8.31b}$$

$$\bar{w}(\underset{\sim}{P},t) = 1 + s\bar{u}(\underset{\sim}{P},t), \tag{8.31c}$$

with

$$u_{ij}(\underset{\sim}{P},t) = \sum_{kl} u_{ij,kl}(\underset{\sim}{P},t)P_{kl}, \tag{8.32a}$$

$$\bar{u}(\underset{\sim}{P},t) = \sum_{ij} u_{ij}(\underset{\sim}{P},t)P_{ij}. \tag{8.32b}$$

Summing (8.13) over j and using (8.14) and (8.31), we obtain

$$\bar{w}(\underset{\sim}{P},t)\Delta p_i = sp_i[u_i(\underset{\sim}{P},t) - \bar{u}(\underset{\sim}{P},t)], \tag{8.33}$$

where

$$p_i u_i(\underset{\sim}{P},t) = \sum_j u_{ij}(\underset{\sim}{P},t)P_{ij}. \tag{8.34}$$

For $t \geq t_1$, (8.30) yields

$$w_{ij,kl}(\underset{\sim}{P},t) = w_{ij,kl}(\underset{\sim}{p}*\underset{\sim}{q},t) + O(s), \tag{8.35a}$$

in which $\underset{\sim}{p}*\underset{\sim}{q}$ indicates evaluation at $P_{ij} = p_i q_j$. From (8.35a) it follows that

$$\bar{w}(\underset{\sim}{P},t) = \bar{w}(\underset{\sim}{p}*\underset{\sim}{q},t) + O(s), \tag{8.35b}$$

where

$$\bar{w}(\underset{\sim}{p}*\underset{\sim}{q},t) = \sum_{ijkl} w_{ij,kl}(\underset{\sim}{p}*\underset{\sim}{q},t)p_i q_j p_k q_l. \tag{8.36}$$

Defining

$$u_{ij}(\underset{\sim}{p}*\underset{\sim}{q},t) = \sum_{kl} u_{ij,kl}(\underset{\sim}{p}*\underset{\sim}{q},t)p_k q_l, \tag{8.37a}$$

$$u_i(\underset{\sim}{p}*\underset{\sim}{q},t) = \sum_j u_{ij}(\underset{\sim}{p}*\underset{\sim}{q},t)q_j, \tag{8.37b}$$

$$\bar{u}(\underset{\sim}{p}*\underset{\sim}{q},t) = \sum_i u_i(\underset{\sim}{p}*\underset{\sim}{q},t)p_i \tag{8.37c}$$

in analogy with (8.32) and (8.34), we deduce for $t \geq t_1$ from (8.33), (8.35b), and analogous relations for u_i and $\bar{u}$ that

$$\bar{w}(\underset{\sim}{p}*\underset{\sim}{q},t)\Delta p_i = sp_i[u_i(\underset{\sim}{p}*\underset{\sim}{q},t) - \bar{u}(\underset{\sim}{p}*\underset{\sim}{q},t)] + O(s^2). \tag{8.38}$$

From (8.38) we infer that on the linkage equilibrium surface the frequency of A_i satisfies

$$\bar{w}(\underset{\sim}{\pi}*\underset{\sim}{\rho},t)\Delta \pi_i = s\pi_i[u_i(\underset{\sim}{\pi}*\underset{\sim}{\rho},t) - \bar{u}(\underset{\sim}{\pi}*\underset{\sim}{\rho},t)]; \tag{8.39}$$

there is a similar equation for p_i.

Comparing (8.38) and (8.39), as well as the analogous equations for q_i and ρ_i, we conclude that for $t \geq t_1$

$$p_i(t) = \pi_i(t) + O(s), \qquad q_i(t) = \rho_i(t) + O(s). \tag{8.40}$$

Inserting (8.40) into (8.30) directly yields the desired result,

$$P_{ij}(t) = \pi_i(t)\rho_j(t) + O(s), \qquad t \geq t_1. \tag{8.41}$$

If the system on the linkage equilibrium surface does not converge to isolated equilibrium points, small deviations due to linkage disequilibrium may greatly alter its final state. Then (8.41) still holds as long as $t_1 \leq t \leq T$, for any fixed time T.

We shall now examine the evolution of the mean fitness. From (8.8b) and (8.12) we have

$$\Delta\bar{w} = \sum_{ijkl} (w'_{ij,kl}P'_{ij}P'_{kl} - w_{ij,kl}P_{ij}P_{kl})$$

$$= \overline{\Delta w} + 2\sum_{ij} w_{ij}\Delta P_{ij} + \sum_{ijkl} w_{ij,kl}\Delta P_{ij}\Delta P_{kl}, \tag{8.42}$$

where

$$\overline{\Delta w} = \sum_{ijkl} \Delta w_{ij,kl}P'_{ij}P'_{kl} \tag{8.43}$$

is the mean of the fitness changes over the next generation. Substituting (8.13) into (8.42) and recalling (8.14) leads to

$$\Delta\bar{w} = \overline{\Delta w} + \bar{w}^{-1}(V_{gam} - 2c\bar{X}) + \bar{w}^{-2}\sum_{ijkl} P_{ij}P_{kl}(w_{ij,kl}-\bar{w})(w_{ij}-\bar{w})(w_{kl}-\bar{w})$$

$$-2c\bar{w}^{-2}\sum_{ij} X_{ij}P_{ij}(w_{ij}-\bar{w}) + c^2\bar{w}^{-2}\sum_{ij} X_{ij}D_{ij}, \tag{8.44}$$

where the *gametic* variance reads

$$V_{gam} = 2\sum_{ij} P_{ij}(w_{ij} - \bar{w})^2, \tag{8.45}$$

and

$$X_{ij} = \sum_{kl} w_{ij,kl}D_{kl} = \sum_{kl} (w_{ij,kl} - \bar{w})D_{kl}, \tag{8.46a}$$

$$\bar{X} = \sum_{ij} P_{ij}X_{ij} = \sum_{kl} w_{kl}D_{kl} = \sum_{kl} (w_{kl} - \bar{w})D_{kl}. \tag{8.46b}$$

Note that (8.44) reduces to the single-locus formula, (4.56), with alleles replaced by gametes, in the absence of recombination (c = 0). Comparing (8.45) and (4.189), we observe that, since the gametes are combined in Hardy-Weinberg proportions, V_{gam} is indeed the additive component of the total genetic variance in the least squares decomposition [cf. (4.184)]

$$V = V_{gam} + V_{gam,dom}, \tag{8.47}$$

where $V_{gam,dom}$ is the variance arising from nonadditivity of the gametic fitnesses. Consequently, if we decompose the gametic variance into its additive (V_g) and epistatic (V_e) components, then V_g will be the additive component of V, and hence may justly be called the genic variance. Kimura (1965) proved this result algebraically for two diallelic loci.

To carry out this analysis of V_{gam}, we set

$$v_{ij} \equiv w_{ij} - \bar{w} = \alpha_i + \beta_j + E_{ij}, \tag{8.48}$$

where α_i and β_j are the average effects of A_i and B_j, and E_{ij} is the epistatic deviation. Minimizing the epistatic variance

$$V_e = 2\sum_{ij} P_{ij}E_{ij}^2 = 2\sum_{ij} P_{ij}(v_{ij} - \alpha_i - \beta_j)^2 \tag{8.49}$$

with respect to the average effects, we obtain

$$\sum_j P_{ij}(v_{ij} - \alpha_i - \beta_j) = \sum_j P_{ij}E_{ij} = 0, \tag{8.50a}$$

$$\sum_i P_{ij}(v_{ij} - \alpha_i - \beta_j) = \sum_i P_{ij}E_{ij} = 0. \tag{8.50b}$$

The average excesses of A_i and B_j are given by

$$p_i a_i = \sum_j P_{ij}v_{ij}, \qquad q_j b_j = \sum_i P_{ij}v_{ij}. \tag{8.51}$$

Hence, (8.48) shows that the means of the average excesses vanish:

$$\sum_i p_i a_i = 0, \qquad \sum_j q_j b_j = 0. \tag{8.52}$$

We rewrite (8.50) in the form

$$p_i a_i = p_i \alpha_i + \sum_j P_{ij}\beta_j, \tag{8.53a}$$

$$q_j b_j = \sum_i P_{ij}\alpha_i + q_j\beta_j, \tag{8.53b}$$

and deduce from (8.52) and (8.53)

$$\sum_i p_i\alpha_i + \sum_j q_j\beta_j = 0. \tag{8.54}$$

Since α_i and β_j occur only in the sum $\alpha_i + \beta_j$, by shifting the appropriate constant between them, owing to (8.54), we can arrange that the average effects have zero means:

$$\sum_i p_i\alpha_i = 0, \qquad \sum_j q_j\beta_j = 0. \tag{8.55}$$

From (8.53) we compute the genic variance

$$V_g = 2\sum_{ij} P_{ij}(\alpha_i + \beta_j)^2 \tag{8.56}$$

$$= 2\sum_{ij} P_{ij}(\alpha_i + \beta_j)\alpha_i + 2\sum_{ij} P_{ij}(\alpha_i + \beta_j)\beta_j$$

$$= 2\sum_i p_i a_i \alpha_i + 2\sum_j q_j b_j \beta_j, \tag{8.57}$$

which manifestly generalizes (4.188) to two loci. Substituting (8.48) into (8.45) and using (8.49), (8.50), and (8.56) proves the additivity property for the gametic variance:

$$V_{gam} = 2\sum_{ij} P_{ij}(\alpha_i + \beta_j + E_{ij})^2$$

$$= V_g + V_e + 4\sum_{ij} P_{ij}(\alpha_i + \beta_j)E_{ij}$$

$$= V_g + V_e. \tag{8.58}$$

Since $D_{ij} = O(s)$ for $t \geq t_1$, (8.58) allows us to rewrite (8.44) as

$$\Delta\bar{w} = V_g + \overline{\Delta w} + V_e - 2c\bar{X} + O(s^3), \qquad t \geq t_1. \tag{8.59}$$

To calculate $V_e - 2c\bar{X}$, we evaluate ΔD_{ij} for $t \geq t_1$, explaining the manipulations below:

$$\Delta D_{ij} = \sum_{kl} (P_{kl}\Delta P_{ij} + P_{ij}\Delta P_{kl} - P_{kj}\Delta P_{il} - P_{il}\Delta P_{kj}) + O(s^2)$$

$$= \sum_{kl} \{P_{kl}P_{ij}[(w_{ij} - \bar{w}) + (w_{kl} - \bar{w})]$$

$$- P_{kj}P_{il}[(w_{il} - \bar{w}) + (w_{kj} - \bar{w})]\} - cD_{ij} + O(s^2)$$

$$= \sum_{kl} p_i q_j p_k q_l [(w_{ij} - \bar{w}) + (w_{kl} - \bar{w}) - (w_{il} - \bar{w}) - (w_{kj} - \bar{w})] - cD_{ij} + O(s^2)$$

$$= \sum_{kl} p_i q_j p_k q_l (E_{ij} + E_{kl} - E_{il} - E_{kj}) - cD_{ij} + O(s^2)$$

$$= \sum_{kl} [P_{kl} P_{ij} (E_{ij} + E_{kl}) - P_{il} P_{kj} (E_{il} + E_{kj})] - cD_{ij} + O(s^2)$$

$$= P_{ij} E_{ij} - cD_{ij} + O(s^2), \qquad t \geq t_1, \tag{8.60}$$

where we employed in the successive equations (8.11), (8.17), (8.23), and the assumption that the fitnesses are approximately constant,

$$\Delta w_{ij,kl} = O(s^2); \tag{8.61}$$

(8.13), (8.14), and (8.17); (8.17) and (8.30); (8.48); (8.17), (8.30), and (8.48); and finally, (8.50). From (8.48) and (8.60) we obtain

$$2 \sum_{ij} E_{ij} \Delta D_{ij} = 2 \sum_{ij} P_{ij} E_{ij}^2 - 2c \sum_{ij} D_{ij} (v_{ij} - \alpha_i - \beta_j) + O(s^3)$$

$$= V_e - 2c\bar{X} + O(s^3), \qquad t \geq t_1, \tag{8.62}$$

where the second step follows from (8.14), (8.46b), (8.48), and (8.49). Inserting (8.62) into (8.59) yields

$$\Delta\bar{w} = V_g + \overline{\Delta w} + 2 \sum_{ij} E_{ij} \Delta D_{ij} + O(s^3), \qquad t \geq t_1. \tag{8.63}$$

Now (8.28) informs us that

$$\Delta\bar{w} = V_g + \overline{\Delta w} + O(s^3), \qquad t \geq t_2, \tag{8.64}$$

as for a single locus (4.59).

Slightly strengthening (8.61) to $\Delta w_{ij,kl} = o(s^2)$ gives the approximate form of the Fundamental Theorem of Natural Selection for two loci, $\Delta\bar{w} \approx V_g$. Therefore, the mean fitness will generally not decrease after the initial adjustment of linkage relations during $t < t_2$. This assertion will be false if the genic variance is particularly small. From (8.13), (8.14), (8.48), and (8.51) we deduce that

$$\bar{w}\Delta p_i = p_i a_i, \qquad \bar{w}\Delta q_j = q_j b_j, \tag{8.65}$$

whence (8.57) becomes

$$V_g = 2\bar{w}\left(\sum_i \alpha_i \Delta p_i + \sum_j \beta_j \Delta q_j\right). \tag{8.66}$$

Hence, $V_g = 0$ if the allelic frequencies are constant, as is true at equilibrium. Thus, the mean fitness is most likely to decrease in the neighborhood of an equilibrium and in special cases in which symmetry conditions dictate the constancy of the gene frequencies. For further discussion, examples of pathological behavior, and other results, the reader may refer to Nagylaki (1977a).

8.3 *Two Diallelic Loci*

The purpose of this section is to present some examples of the properties of two diallelic loci. Karlin (1975) gives many other results in his extensive review. It will be convenient to simplify the notation of Section 8.2. Let x_1, x_2, x_3, x_4, represent the frequencies of the gametes *AB*, *Ab*, *aB*, *ab*, as in Section 3.3. If w_{ij} denotes the fitness of the genotype composed of gametes i and j (i,j = 1, 2, 3, 4), the gametic fitnesses and the mean fitness read

$$w_i = \sum_j w_{ij}x_j, \qquad \bar{w} = \sum_{ij} w_{ij}x_ix_j. \tag{8.67}$$

It is easy to show that (8.10) and (8.11) become (Problem 8.4)

$$\bar{w}\Delta x_i = x_i(w_i - \bar{w}) - \varepsilon_i cD, \tag{8.68}$$

where $\varepsilon_1 = \varepsilon_4 = 1$, $\varepsilon_2 = \varepsilon_3 = -1$, and

$$D_{11} = D_{22} = -D_{12} = -D_{21} \equiv D = w_{14}x_1x_4 - w_{23}x_2x_3. \tag{8.69}$$

1. The increase in frequency of a rare allele:

Suppose the allele *a* is fixed and linked to an overdominant polymorphism at the *B*-locus. Thus, (4.29) tells us that the equilibrium frequencies of the gametes *aB* and *ab* are

$$\hat{u} = (w_{34} - w_{44})/(2w_{34} - w_{33} - w_{44})$$

and $\hat{v} = 1 - \hat{u}$. The equilibrium fitness of the genotype *aa*-- is just

$$\hat{\bar{w}} = \hat{u}^2 w_{33} + 2\hat{u}\hat{v}w_{34} + \hat{v}^2 w_{44},$$

the mean fitness of the population at equilibrium.

If a rare new mutant *A* appears in the population, its frequency will increase if the two-locus polymorphism is unstable. Since the *Bb* polymorphism is stable in the absence of *A*, the equilibria with the gametes *aB* or *ab* fixed are evidently unstable. Therefore, if the introduction of *A* makes the *Bb* polymorphism unstable, *A* will not only increase initially in frequency, but (possibly excluding some very special initial conditions) it will be permanently established in the population. As explained in Section 5.5, two-locus instability will occur if at least one eigenvalue of the linearized system exceeds unity in modulus. We reproduce below the results of Bodmer and Felsenstein (1967).

The fitnesses of the gametes *AB* and *Ab* at equilibrium (i.e., as their frequencies tend to zero) read

$$\hat{w}_1 = \hat{u}w_{13} + \hat{v}w_{14}, \qquad \hat{w}_2 = \hat{u}w_{23} + \hat{v}w_{24}.$$

The allele *A* has frequency p and fitness $p^{-1}(x_1w_1 + x_2w_2)$. In linkage equilibrium, $x_1 = pu$ and $x_2 = pv$. Therefore, we may interpret

$$\hat{w}_A = \hat{u}\hat{w}_1 + \hat{v}\hat{w}_2$$

as the stationary fitness of *A* at linkage equilibrium. We define the critical value

$$c^* = \left| \frac{(\hat{\bar{w}} - \hat{w}_1)(\hat{\bar{w}} - \hat{w}_2)}{\hat{u}w_{23}(\hat{\bar{w}} - \hat{w}_1) + \hat{v}w_{14}(\hat{\bar{w}} - \hat{w}_2)} \right|$$

of the cross-over fraction. Bodmer and Felsenstein (1967) obtained the following conditions for the increase of *A*.

(i) $\hat{w}_1 = \hat{w}_2$:

Here $\hat{w}_A = \hat{w}_1 = \hat{w}_2$, and the frequency of *A* increases if $\hat{w}_A > \hat{\bar{w}}$, i.e., roughly, the frequency of the new mutant increases if it is fitter than the original allele, a.

(ii) $\hat{w}_1 > \hat{w}_2$:

(a) $\hat{w}_1 > \hat{w}_2 > \hat{\bar{w}}$:

Both gametes carrying *A* are fitter than the population, so *A* spreads.

(b) $\hat{\bar{w}} > \hat{w}_1 > \hat{w}_2$:

Since *A* decreases the fitness of every gamete, it is lost.

(c) $\hat{w}_1 > \hat{\bar{w}} \geq \hat{w}_2$:

(α) $\hat{\bar{w}}(\hat{u}w_{23} + \hat{v}w_{14}) \leq \hat{u}w_{23}\hat{w}_1 + \hat{v}w_{14}\hat{w}_2$: *A* increases.

(β) $\hat{\bar{w}}(\hat{u}w_{23} + \hat{v}w_{14}) > \hat{u}w_{23}\hat{w}_1 + \hat{v}w_{14}\hat{w}_2$: *A* increases if $c < c^*$.

(iii) $\hat{w}_1 < \hat{w}_2$:

This is the same as Case (ii) with the interchange $B \leftrightarrow b$

throughout.

If there is no position effect, then $w_{14} = w_{23}$, and the most interesting case, (iic), simplifies to the following two situations.

(α) $\hat{\bar{w}} \leq \hat{w}_A$:

The frequency of A increases.

(β) $\hat{\bar{w}} > \hat{w}_A$:

A will spread if $c < c^*$. Thus, for sufficiently tight linkage, the frequency of the less fit allele can increase. We saw at the end of Section 4.3 that this was not possible for fitness dependent only on a single locus. Note that for weak selection $c^* = O(s)$.

2. The increase in frequency of two rare alleles:

We assume the gamete ab is fixed, and the alleles A and B are introduced at low frequencies. If the equilibrium $x_4 = 1$ is stable, A and B will be lost. If it is unstable, though the frequency of one of the alleles may decrease temporarily (Bodmer and Felsenstein, 1967), in general, both will become more common after a short time. Whether one of the alleles is ultimately lost cannot be determined by local analysis.

The linearized recursion relation satisfied by the vector

$$\underset{\sim}{x} = \begin{pmatrix} x_1 \\ x_2 \\ x_3 \end{pmatrix}$$

reads (Problem 8.4) $\underset{\sim}{x}' = C\underset{\sim}{x}$, with

$$C = \begin{pmatrix} \lambda_1 & 0 & 0 \\ \alpha & \lambda_2 & 0 \\ \alpha & 0 & \lambda_3 \end{pmatrix}, \tag{8.70a}$$

in which $\alpha = cw_{14} / w_{44}$, and the eigenvalues of C are

$$\lambda_1 = (1-c)w_{14}/w_{44}, \qquad \lambda_2 = w_{24}/w_{44}, \qquad \lambda_3 = w_{34}/w_{44}. \tag{8.70b}$$

The equilibrium is stable if all the eigenvalues are less than one. If $\lambda_2, \lambda_3 < 1$, we have instability for sufficiently tight linkage, $c < 1 - (w_{44}/w_{14}) \equiv \tilde{c}$. Thus, linkage matters if Ab and aB are less fit at equilibrium than ab ($w_{24}, w_{34} < w_{44}$) and the equilibrium fitness of AB is neither too small nor too large, $w_{44} < w_{14} \leq 2w_{44}$ ($0 < \tilde{c} \leq 1/2$).

Observe that for weak selection $\tilde{c} = O(s)$.

3. The additive and multiplicative models:

The results given below, as well as many more concerning other models, are presented by Karlin (1975), who gives references to the pertinent original papers. Assuming the absence of a position effect, $w_{14} = w_{23}$, we can use loci instead of gametes to construct a 3×3 fitness matrix instead of our former 4×4 symmetric one:

$$W = (W_{ij}) = \begin{matrix} \\ AA \\ Aa \\ aa \end{matrix} \begin{matrix} BB \quad Bb \quad bb \\ \begin{pmatrix} w_{11} & w_{12} & w_{22} \\ w_{13} & w_{14} & w_{24} \\ w_{33} & w_{34} & w_{44} \end{pmatrix} \end{matrix},$$

$i,j = 1,2,3$. For instance, $W_{22} = w_{14} = w_{23} \neq w_{22}$. The contributions of the individual loci to the genotypic fitnesses are

AA	Aa	aa	BB	Bb	bb
α_1	α_2	α_3	β_1	β_2	β_3 .

Define

$$\hat{x}_1 = \hat{p}\hat{u}, \qquad \hat{x}_2 = \hat{p}\hat{v}, \qquad \hat{x}_3 = \hat{q}\hat{u}, \qquad \hat{x}_4 = \hat{q}\hat{v}, \tag{8.71a}$$

where the equilibrium frequencies of A and B read

$$\hat{p} = (\alpha_2 - \alpha_3)/(2\alpha_2 - \alpha_1 - \alpha_3), \qquad \hat{u} = (\beta_2 - \beta_3)/(2\beta_2 - \beta_1 - \beta_3). \tag{8.71b}$$

Clearly, the equilibrium (8.71) can exist only if there is either over- or underdominance at each locus. Evidently, $D = 0$ at this equilibrium.

(i) Additive loci:

With no additive epistasis, we have $W_{ij} = \alpha_i + \beta_j$, and the following results hold.

(a) An internal equilibrium (i.e., one with $x_i > 0$ for all i) exists only if there is either over- or underdominance at each locus, in which case (8.71) is the unique internal equilibrium.

(b) The equilibrium (8.71) is globally asymptotically stable if and only if there is overdominance at each locus, i.e.,

$$\alpha_2 > \alpha_1, \alpha_3, \qquad \beta_2 > \beta_1, \beta_3. \tag{8.72}$$

(c) The mean fitness is nondecreasing, the change in mean fitness being zero only at equilibrium.

(d) If only one of the conditions (8.72) holds, say the first, the frequency of A converges globally to $\hat{p}$, while B or b will be lost.

(e) Even with multiple alleles at each locus, there exists at most one internal equilibrium, and it is globally stable if and only if it is locally stable.

(ii) Multiplicative loci:

In the absence of multiplicative epistasis, we have $W_{ij} = \alpha_i \beta_j$, and the system has the following properties.

(a) With either over- or underdominance at each locus, the equilibrium (8.71) exists.

(b) If (8.72) holds, the equilibrium (8.71) is locally stable if and only if the recombination frequency, c, exceeds a critical value, c_0, i.e., with overdominance at each locus, the expected $D = 0$ equilibrium is stable if and only if linkage is sufficiently loose. From the explicit expression for c_0 (Karlin, 1975), it is easy to see that for weak selection $c_0 = O(s^2)$, so that the condition $c > c_0$ is not very stringent.

(c) If (8.72) applies and $c < c_0$, there exist exactly two locally stable internal equilibria, one with $D > 0$ and one with $D < 0$. Since the quadratic $\bar{w}$ can have at most one internal maximum, the mean fitness can decrease in the neighborhood of at least one of the two stable equilibria. The discussion at the end of Section 8.2 is pertinent here.

(d) The equilibrium (8.71) is not a local maximum of $\bar{w}$. Hence, $\bar{w}$ can decrease in the neighborhood of (8.71) when that equilibrium is stable. Refer again to the end of the previous section.

(e) The sign of the linkage disequilibrium D is invariant. Therefore, by continuity, $D = 0$ implies $D' = 0$. Thus, depending on its initial state, a population remains forever in one of the regions $D > 0$ or $D < 0$, or on the surface $D = 0$.

(f) The mean fitness is nondecreasing on the $D = 0$ surface. When (8.71) exists, it is a local maximum of $\bar{w}$ on that surface.

(g) There exist at most two stable equilibria (internal and boundary).

8.4 *Continuous Model with Overlapping Generations*

We shall now discuss the behavior of the multiallelic two-locus system in continuous time. We shall assume that either there is no age structure or a stable age distribution has been reached, and therefore commence with the two-locus generalization of (4.131), in which time is the only independent variable. For two alleles at each locus, the formulation and analysis to follow have been carried out in Nagylaki and Crow (1974) and Nagylaki (1976).

Our basic variables are the genotypic frequencies. Let $S_{ij,kl}(t)$ represent the frequency of $A_iB_j\,/\,A_kB_l$ individuals at time t. From this we compute the frequencies of the gamete A_iB_j and the alleles A_i and B_j:

$$P_{ij} = \sum_{kl} S_{ij,kl}, \qquad p_i = \sum_j P_{ij}, \qquad q_j = \sum_i P_{ij}. \tag{8.73}$$

We denote the number of A_iB_j/A_kB_l individuals by $n_{ij,kl}$ and the total population size by N. Hence,

$$S_{ij,kl} = n_{ij,kl}\,/\,N. \tag{8.74}$$

We designate the frequency of matings between $A_\alpha B_\beta\,/\,A_\gamma B_\delta$ and $A_\varepsilon B_\zeta\,/\,A_\eta B_\theta$ by $X_{\alpha\beta,\gamma\delta;\varepsilon\zeta,\eta\theta}$, and the fertility of such a union, normalized relative to individuals, as in (4.130), by $f_{\alpha\beta,\gamma\delta;\varepsilon\zeta,\eta\theta}$. $R_{ij;\alpha\beta,\gamma\delta}$, given by (8.9), is the probability of drawing an A_iB_j gamete from an $A_\alpha B_\beta\,/\,A_\gamma B_\delta$ individual.

The differential equation for the genotypic numbers reads

$$\dot{n}_{ij,kl} = N \sum X_{\alpha\beta,\gamma\delta;\varepsilon\zeta,\eta\theta} f_{\alpha\beta,\gamma\delta;\varepsilon\zeta,\eta\theta} R_{ij;\alpha\beta,\gamma\delta} R_{kl;\varepsilon\zeta,\eta\theta} - d_{ij,kl} n_{ij,kl}, \tag{8.75}$$

where $d_{ij,kl}$ is the death rate of $A_iB_j\,/\,A_kB_l$ individuals. To simplify the notation, in (8.75) we introduced the convention that unspecified summations run over all indices not appearing on the left-hand side of the equation. We define the average mortality, fertility, and fitness by

$$\bar{d} = \sum d_{ij,kl} S_{ij,kl}, \tag{8.76a}$$

$$\bar{f} = \sum X_{\alpha\beta,\gamma\delta;\varepsilon\zeta,\eta\theta} f_{\alpha\beta,\gamma\delta;\varepsilon\zeta,\eta\theta}, \tag{8.76b}$$

$$\bar{m} = \bar{f} - \bar{d}. \tag{8.76c}$$

The recombination functions yield the Mendelian single-locus formulae [see (8.9)]

$$\sum_j R_{ij;kl,mn} = \tfrac{1}{2}(\delta_{ik} + \delta_{im}), \qquad \sum_i R_{ij;kl,mn} = \tfrac{1}{2}(\delta_{jl} + \delta_{jn}), \tag{8.77a}$$

and the normalization

$$\sum_{ij} R_{ij;kl,mn} = 1. \tag{8.77b}$$

From (8.75), (8.76), and (8.77b) we obtain

$$\dot{N} = \bar{m}N. \tag{8.78}$$

Differentiating (8.74) and substituting (8.75) and (8.78) gives our fundamental equations,

$$\dot{S}_{ij,kl} = \sum X_{\alpha\beta,\gamma\delta;\varepsilon\zeta,\eta\theta} f_{\alpha\beta,\gamma\delta;\varepsilon\zeta,\eta\theta} R_{ij;\alpha\beta,\gamma\delta} R_{kl;\varepsilon\zeta,\eta\theta}$$

$$-(d_{ij,kl} + \bar{m})S_{ij,kl}. \tag{8.79}$$

The evolution of the population is determined by (8.79) as soon as the mating frequencies, fecundities, and mortalities are specified as functions of time and the genotypic proportions. The average death rate of an individual carrying the gamete A_iB_j is d_{ij}, where

$$P_{ij}d_{ij} = \sum_{kl} S_{ij,kl}d_{ij,kl}. \tag{8.80}$$

From (8.73), (8.77b), (8.79), and (8.80) we deduce

$$\dot{P}_{ij} = \sum X_{\alpha\beta,\gamma\delta;\varepsilon\zeta,\eta\theta} f_{\alpha\beta,\gamma\delta;\varepsilon\zeta,\eta\theta} R_{ij;\alpha\beta,\gamma\delta} - (d_{ij} + \bar{m})P_{ij}. \tag{8.81}$$

We define the rate at which an A_iB_j / A_kB_l individual gives birth, $b_{ij,kl}$, by

$$S_{ij,kl}b_{ij,kl} = \sum X_{ij,kl;\alpha\beta,\gamma\delta} f_{ij,kl;\alpha\beta,\gamma\delta}. \tag{8.82a}$$

Then the fertilities of individuals carrying A_iB_j and of those carrying A_i are given by

$$P_{ij}b_{ij} = \sum_{kl} S_{ij,kl}b_{ij,kl}, \tag{8.82b}$$

$$p_ib_i = \sum_j P_{ij}b_{ij}. \tag{8.82c}$$

Of course, the allelic mortality of A_i satisfies

$$p_i d_i = \sum_j P_{ij} d_{ij}, \tag{8.83}$$

and

$$m_i = b_i - d_i, \tag{8.84}$$

is the Malthusian parameter of A_i. Then (8.73), (8.77a), (8.81), (8.82), (8.83), and (8.84) lead easily to the usual differential equation,

$$\dot{p}_i = p_i(m_i - \bar{m}), \tag{8.85}$$

for the frequency of A_i. Evidently, q_j satisfies a similar equation.

As pointed out in Sections 4.1 and 4.10, equations for marginal frequencies like (8.81) and (8.85), though highly informative, depend on genotypic proportions, and are consequently incomplete. By inserting (8.9), it is trivial to display explicitly in (8.79) and (8.81) the terms corresponding to recombination in 0, 1, or 2 individuals in the mated pair. Since the ensuing analysis will be restricted to weak selection, we shall save writing by decomposing (8.79) and (8.81) only in that case. We suppose henceforth that mating is random:

$$X_{\alpha\beta,\gamma\delta;\varepsilon\zeta,\eta\theta} = S_{\alpha\beta,\gamma\delta} S_{\varepsilon\zeta,\eta\theta}. \tag{8.86}$$

This reduces (8.82a) to

$$b_{ij,kl} = \sum S_{\alpha\beta,\gamma\delta} f_{ij,kl;\alpha\beta,\gamma\delta}.$$

Let us now study slow selection:

$$f_{\alpha\beta,\gamma\delta;\varepsilon\zeta,\eta\theta} = b + O(s), \qquad d_{ij,kl} = d + O(s), \tag{8.87}$$

where b is a constant and s is the intensity of selection. If time is measured in generations, b will usually be close to unity (as will d). Inserting (8.9) and (8.87) into (8.79) yields

$$\dot{S}_{ij,kl} = b\{(1-c)^2 P_{ij} P_{kl} + c(1-c)[P_{ij} \sum S_{k\zeta,\eta l} + P_{kl} \sum S_{i\beta,\gamma j}]$$
$$+ c^2 \sum S_{i\beta,\gamma j} S_{k\zeta,\eta l} - S_{ij,kl}\} + O(s). \tag{8.88}$$

Hence,

$$\dot{P}_{ij} = -bcD_{ij} + O(s), \tag{8.89}$$

with the linkage disequilibrium

$$D_{ij} = \sum_{kl} (S_{ij,kl} - S_{il,kj}) = P_{ij} - \sum_{kl} S_{il,kj}. \tag{8.90}$$

Differentiating (8.90) and substituting (8.88), (8.89), and then

$$\sum_{kl} S_{il,kj} = P_{ij} - D_{ij}, \tag{8.91}$$

we find

$$\dot{D}_{ij} = -b(1+c)D_{ij} + bD^r_{ij} + O(s), \tag{8.92}$$

where

$$D^r_{ij} = P_{ij} - p_i q_j \tag{8.93}$$

would be the linkage disequilibrium if the population were in Hardy-Weinberg proportions. Note that for weak selection (8.11) and (8.93) agree. From (8.85) and the corresponding equation for q_j we infer

$$\dot{p}_i = O(s), \qquad \dot{q}_j = O(s). \tag{8.94}$$

Using (8.89) and (8.94) in the derivative of (8.93), we obtain

$$\dot{D}^r_{ij} = -bcD_{ij} + O(s). \tag{8.95}$$

We shall also require

$$Q_{ij,kl} = S_{ij,kl} - P_{ij}P_{kl} \tag{8.96}$$

as a measure of the deviation of gametes from Hardy-Weinberg proportions. This is the natural generalization of (4.149). Differentiating (8.96) and inserting (8.88), (8.89), and (8.91) yields

$$\dot{Q}_{ij,kl} = -bQ_{ij,kl} + bc^2 D_{ij}D_{kl} + O(s). \tag{8.97}$$

Concentrating our attention first on the linkage disequilibria, we rewrite (8.92) and (8.95) in matrix form:

$$\underset{\sim}{D}_{ij} = \begin{pmatrix} D_{ij} \\ D^r_{ij} \end{pmatrix}, \qquad C = b\begin{pmatrix} 1+c & -1 \\ c & 0 \end{pmatrix}, \tag{8.98a}$$

$$\dot{\underset{\sim}{D}}_{ij} = -C\underset{\sim}{D}_{ij} + s\underset{\sim}{g}_{ij}(S,t). \tag{8.98b}$$

Here $\underset{\sim}{g}_{ij}$ is a complicated function of genotypic frequencies and possibly of time. By the variation-of-parameters (see, e.g., Brauer and Nohel, 1969, p. 72)

$$\underset{\sim}{D}_{ij}(t) = e^{-Ct}\underset{\sim}{D}_{ij}(0) + s\int_0^t e^{-(t-\tau)C}\underset{\sim}{g}_{ij}[S(\tau),\tau]d\tau. \tag{8.99}$$

From (8.98a) we compute the eigenvalues,

$$\lambda_1 = b, \qquad \lambda_2 = bc, \tag{8.100a}$$

and eigenvectors

$$\underset{\sim}{V}_1 = \begin{pmatrix} 1 \\ c \end{pmatrix}, \qquad \underset{\sim}{V}_2 = \begin{pmatrix} 1 \\ 1 \end{pmatrix}, \tag{8.100b}$$

of C. Then the method employed in Section 5.1 produces a more explicit form of (8.99), viz.,

$$D_{ij}(t) = (1-c)^{-1}\{[D_{ij}(0)-D^r_{ij}(0)]e^{-bt}+[D^r_{ij}(0)-cD_{ij}(0)]e^{-bct}\} +O(s), \tag{8.101a}$$

$$D^r_{ij}(t) = (1-c)^{-1}\{c[D_{ij}(0)-D^r_{ij}(0)]e^{-bt}+[D^r_{ij}(0)-cD_{ij}(0)]e^{-bct}\} +O(s). \tag{8.101b}$$

From (8.97) we find in the same manner

$$Q_{ij,kl}(t) = e^{-bt}Q_{ij,kl}(0) + bc^2e^{-bt}\int_0^t e^{b\tau}D_{ij}(\tau)D_{kl}(\tau)d\tau + O(s). \tag{8.102}$$

In the neutral case ($s = 0$), the gene frequencies are constant, so we can explicitly calculate the gametic frequencies from (8.93) and (8.101b). Substituting (8.101a) into (8.102) and performing the elementary integration, we get $Q_{ij,kl}$, with which we deduce the genotypic proportions from (8.96). Observe that the deviations from linkage equilibrium and Hardy-Weinberg proportions approach zero exponentially. For the linkage disequilibria to remain zero as the system evolves, their initial values must be zero, i.e., for $D_{ij}(t) = 0$ or $D^r_{ij}(t) = 0$ we must have $D_{ij}(0) = D^r_{ij}(0) = 0$. In that case, the integral in (8.102) is absent, and (8.102) reduces to the single-locus formula. Note that even if the population is initially in Hardy-Weinberg proportions, linkage disequilibrium will generate departures from Hardy-Weinberg ratios, i.e., $D_{ij}(t) \neq 0$ implies $Q_{ij,kl}(t) \neq 0$, $t > 0$.

Returning to slow selection, (8.101) allows us to infer that

$$\underset{\sim}{D}_{ij}(t) = O(s), \qquad t \geq t_1, \tag{8.103}$$

where $t_1 \approx -(bc)^{-1}\ln s$. For loose linkage, t_1 is about 5 or 10 generations, as in Section 5.2, and for tight linkage ($s \ll c \ll 1$), it is the same as the corresponding time with discrete nonoverlapping generations. The fact that

$$Q_{ij,kl}(t) = O(s), \qquad t \geq t_1, \tag{8.104}$$

is brought out by decomposing the integral in (8.102) into parts from 0 to t_1 and from t_1 to t and using (8.103). Thus, the rapid approach to nearly random combination we found for a single locus in continuous time and for two loci with discrete nonoverlapping generations still applies here.

To demonstrate that the deviations from random combination tend to approximate constancy, we put $\underset{\sim}{D}_{ij} = s\underset{\sim}{D}^0_{ij}$ and $Q_{ij,kl} = sQ^0_{ij,kl}$, and express (8.97) and (8.98b) for $t \geq t_1$ in the form

$$\dot{Q}^0_{ij,kl} = -bQ^0_{ij,kl} + h_{ij,kl}(S,t), \tag{8.105a}$$

$$\underset{\sim}{D}^0_{ij} = -c\underset{\sim}{D}^0_{ij} + \underset{\sim}{g}_{ij}(S,t), \tag{8.105b}$$

where h_{ij} comes from the $O(s)$ term in (8.97). Equation (8.105a) corresponds precisely to (4.156). We suppose

$$\frac{\partial \underset{\sim}{g}_{ij}}{\partial t} = O(s), \qquad \frac{\partial h_{ij,kl}}{\partial t} = O(s), \tag{8.106}$$

and rewrite (8.105b) as

$$\frac{d}{dt}[\underset{\sim}{D}^0_{ij} - c^{-1}\underset{\sim}{g}_{ij}(S,t)] = -c[\underset{\sim}{D}^0_{ij} - c^{-1}\underset{\sim}{g}_{ij}(S,t)] + O(s). \tag{8.107}$$

Therefore, exactly as in (8.99), we conclude

$$\underset{\sim}{D}^0_{ij}(t) = c^{-1}\underset{\sim}{g}_{ij}(S,t) + O(s) \tag{8.108}$$

for $t \geq t_2 \approx 2t_1$. From (8.93), (8.94), (8.96), (8.103), and (8.104) we obtain

$$\dot{S}_{ij,kl}(t) = O(s), \qquad t \geq t_1. \tag{8.109}$$

With the mild assumption, (8.106), that the explicit time dependence of the fertilities and mortalities is of $O(s^2)$, and the aid of (8.108) and (8.109), we obtain $\dot{\underset{\sim}{D}}^0_{ij}(t) = O(s)$ for $t \geq t_2$. Hence

$$\dot{\underset{\sim}{D}}_{ij}(t) = O(s^2), \qquad \dot{Q}_{ij,kl}(t) = O(s^2), \qquad t \geq t_2. \tag{8.110}$$

By virtue of (8.103) and (8.104), for $t \geq t_1$ the exact system satisfying (8.88) is within $O(s)$ of the much simpler system in Hardy-Weinberg proportions and linkage equilibrium. Since the proof of this statement involves little more than a Taylor expansion, the reader will have no trouble generalizing the demonstration in Section 4.10 to two loci.

Finally, we wish to consider the behavior of the mean fitness. Kimura (1958) calculated the rate of change of the mean fitness by decomposing the genotypic fitnesses first with respect to loci and then with respect to alleles within loci. It will be closer to our approach to first do a least squares decomposition relative to gametes, as was done in effect in Section 4.10, and then to follow Section 8.2 in a least squares decomposition with respect to alleles within gametes. One finds (Problem 8.8)

$$\dot{\bar{m}} = V_g + \bar{\dot{m}} + 2\overline{\overset{\circ}{\phi}E} + \overline{\overset{\circ}{\Theta}\Delta}, \tag{8.111}$$

where V_g is the genic variance,

$$\bar{\dot{m}} = \sum_{ijkl} \dot{m}_{ij,kl} S_{ij,kl},$$

the circle indicates the logarithmic derivative, as below (4.191),

$$\phi_{ij} = P_{ij} / p_i q_j, \qquad \Theta_{ij,kl} = S_{ij,kl} / P_{ij} P_{kl}, \tag{8.112}$$

E_{ij} and $\Delta_{ij,kl}$ denote the epistatic deviations within the least squares gametic fitnesses (which, owing to the deviation from Hardy-Weinberg proportions, are not the same as the marginal gametic fitnesses m_{ij}) and the "dominance" deviations between gametes, and

$$\overline{\overset{\circ}{\phi}E} = \sum_{ij} \overset{\circ}{\phi}_{ij} E_{ij} P_{ij},$$

$$\overline{\overset{\circ}{\Theta}\Delta} = \sum_{ijkl} \overset{\circ}{\Theta}_{ij,kl} \Delta_{ij,kl} S_{ij,kl}.$$

Suppose $\bar{\dot{m}} = o(s^2)$, i.e., the genotypic fitnesses are almost constant. Since (8.93), (8.96), and (8.112) give

$$\phi_{ij} = 1 + D^r_{ij} / (p_i q_j), \qquad \Theta_{ij,kl} = 1 + Q_{ij,kl} / (P_{ij} P_{kl}),$$

(8.109) and (8.110) imply

$$\overset{\circ}{\phi}_{ij}(t) = O(s^2), \qquad \overset{\circ}{\Theta}_{ij,kl}(t) = O(s^2)$$

for $t \geq t_2$. Therefore, the last two terms in (8.111) are of $O(s^3)$, and for $t \geq t_2$ we have the approximate Fundamental Theorem of Natural Selection, $\dot{\bar{m}} \approx V_g$.

8.5 *Problems*

8.1. For the model (8.10), show that an equilibrium is a stationary point of the mean fitness if and only if the linkage disequilibria vanish there. Moran (1967) proved this for two diallelic loci.

8.2. Assume that fitnesses are multiplicative in (8.10): $w_{ij,kl} = \alpha_{ik}\beta_{jl}$, where α_{ik} $(= \alpha_{ki})$ and β_{jl} $(= \beta_{lj})$ are the contributions of A_iA_k and B_jB_l. Notice that with multiplicative fitnesses there is no position effect. Suppose the population is initially in linkage equilibrium: $P_{ij}(0) = p_i(0)q_j(0)$. Prove that

(a) it remains in linkage equilibrium: $P_{ij}(t) = p_i(t)q_j(t)$;

(b) the gene frequencies evolve independently at the two loci according to the single-locus equations

$$p_i' = p_i\alpha_i / \bar{\alpha}, \qquad q_j' = q_j\beta_j / \bar{\beta}, \tag{8.113a}$$

$$\alpha_i = \sum_k \alpha_{ik}p_k, \qquad \beta_j = \sum_l \beta_{jl}q_l, \tag{8.113b}$$

$$\bar{\alpha} = \sum_i \alpha_i p_i, \qquad \bar{\beta} = \sum_j \beta_j q_j; \tag{8.113c}$$

(c) the mean fitness is $\bar{w} = \bar{\alpha}\bar{\beta}$, and consequently cannot decrease.

Observe that (a) generalizes the preservation of $D = 0$ in the diallelic model, (b) includes the existence of the over- and underdominant equilibria (8.71) in that model, and (c) generalizes $\Delta\bar{w} \geq 0$ on the $D = 0$ surface with two alleles.

8.3. Assume that the fitnesses in (8.10) are additive: $w_{ij,kl} = \alpha_{ik} + \beta_{jl}$, where α_{ik} $(= \alpha_{ki})$ and β_{jl} $(= \beta_{lj})$ are the contributions of A_iA_k and B_jB_l. Then there is no position effect.

(a) Show that the gene frequencies satisfy

$$\bar{w}p_i' = p_i\alpha_i + \sum_j \beta_j P_{ij},$$

$$\bar{w}q_j' = q_j\beta_j + \sum_i \alpha_i P_{ij},$$

where α_i, β_j, $\bar{\alpha}$, and $\bar{\beta}$ are given by (8.113b) and (8.113c), and $\bar{w} = \bar{\alpha} + \bar{\beta}$. Observe that the mean fitness depends only on the allelic frequencies, and, if the gametic frequencies,

P_{ij}, are specified, the allelic frequencies in the next generation, p_i' and q_j', are independent of the cross-over fraction, c. Therefore, p_i' and q_j', for given P_{ij}, are the same as in the single-locus situation, $c = 0$. But then the same holds for $\overline{w}$, showing that the mean fitness is nondecreasing (Ewens, 1969a).

(b) Prove that linkage equilibrium is generally not preserved.

(c) Demonstrate that if

$$P_{ij} = p_i q_j, \qquad \alpha_i = \overline{\alpha}, \qquad \beta_j = \overline{\beta},$$

the population is in equilibrium. This generalizes the existence of (8.71) in the two-allele model.

8.4. Derive (8.68) and (8.69).

8.5. Deduce (8.70).

8.6. Prove that the epistatic parameters

$$E_i = \sum_j \varepsilon_j w_{ij}$$

vanish for the diallelic model (8.68) if the fitnesses are additive between loci.

8.7. Demonstrate that the multiplicative epistatic parameters

$$e_i = \sum_j \varepsilon_j \ln w_{ij}$$

vanish for the diallelic model (8.68) if the fitnesses are multiplicative between loci.

8.8. Derive (8.100) and (8.101).

8.9. Prove (8.111).

REFERENCES

Atkinson, F.V., Watterson, G.A., and Moran, P.A.P. 1960. A matrix inequality. *Quart. J. Math.* **11**, 137-140.

Baum, L.E., and Eagon, J.A. 1967. An inequality with applications to statistical estimation for probabilistic functions of Markov processes and to a model for ecology. *Bull. Am. Math. Soc.* **73**, 360-363.

Bellman, R. 1949. *A Survey of the Theory of the Boundedness, Stability, and Asymptotic Behavior of Solutions of Linear and Non-linear Differential and Difference Equations.* Office of Naval Research, Washington, D.C.

Bennett, J.H. 1954. On the theory of random mating. *Ann. Eugen.* **18**, 311-317.

Bennett, J.H. 1957. Selectively balanced polymorphism at a sex-linked locus. *Nature* **180**, 1363-1364.

Bodmer, W.F. 1965. Differential fertility in population genetics models. *Genetics* **51**, 411-424.

Bodmer, W.F., and Cavalli-Sforza, L.L. 1968. A migration matrix model for the study of random genetic drift. *Genetics* **59**, 565-592.

Brauer, F., and Nohel, J.A. 1969. *The Qualitative Theory of Ordinary Differential Equations.* W.A. Benjamin, New York.

Bulmer, M.G. 1972. Multiple niche polymorphism. *Am. Nat.* **106**, 254-257.

Cannings, C. 1967. Equilibrium, convergence, and stability at a sex-linked locus under natural selection. *Genetics* **56**, 613-617.

Cannings, C. 1968. Fertility difference between homogamous and heterogamous matings. *Genet. Res.* **11**, 289-301.

Cannings, C. 1968a. Equilibrium under selection at a multi-allelic sex-linked locus. *Biometrics* **11**, 187-189.

Cannings, C. 1971. Natural selection at a multiallelic autosomal locus with multiple niches. *J. Genet.* **60**, 255-259.

Cavalli-Sforza, L.L., and Bodmer, W.F. 1971. *The Genetics of Human Populations.* W.H. Freeman, San Francisco.

Charlesworth, B. 1970. Selection in populations with overlapping generations. I. The use of Malthusian parameters in population genetics. *Theor. Pop. Biol.* **1**, 352-370.

Charlesworth, B. 1974. Selection in populations with overlapping generations. VI. Rates of change of gene frequency and population

growth rate. *Theor. Pop. Biol.* 6, 108-133.
Christiansen, F.B. 1974. Sufficient condition for protected polymorphism in a subdivided population. *Am. Nat.* 108, 157-166.
Christiansen, F.B. 1975. Hard and soft selection in a subdivided population. *Am. Nat.* 109, 11-16.
Conley, C.C. 1972. Unpublished University of Wisconsin lecture notes.
Cornette, J.L. 1975. Some basic elements of continuous selection models. *Theor. Pop. Biol.* 8, 301-313.
Cotterman, C.W. 1937. Indication of unit factor inheritance in data comprising but a single generation. *Ohio J. Sci.* 37, 75-81.
Crow, J.F. 1976. *Genetics Notes*. 7th ed. Burgess, Minneapolis.
Crow, J.F., and Kimura, M. 1970. *An Introduction to Population Genetics Theory*. Harper and Row, New York.
Crow, J.F., and Nagylaki, T. 1976. The rate of change of a character correlated with fitness. *Am. Nat.* 110, 207-213, 400.
Deakin, M.A.B. 1966. Sufficient conditions for genetic polymorphism. *Am. Nat.* 100, 690-692.
de Finetti, B. 1926. Considerazioni matematiche sul l'ereditariete mendeliana. *Metron* 6, 1-41.
Dempster, E.R. 1955. Maintenance of genetic heterogeneity. *Cold Spring Harbor Symp. Quant. Biol.* 20, 25-32.
Ewens, W.J. 1969. *Population Genetics*. Methuen, London.
Ewens, W.J. 1969a. A generalized Fundamental Theorem of Natural Selection. *Genetics* 63, 531-537.
Falk, C.T., and Li, C.C. 1969. Negative assortative mating: exact solution to a simple model. *Genetics* 62, 215-223.
Felsenstein, J., and Taylor, B. 1974. *A Bibliography of Theoretical Population Genetics*. National Technical Information Service, U.S. Department of Commerce, Springfield, Va.
Finney, D.J. 1952. The equilibrium of a self-incompatible polymorphic species. *Genetica* 26, 33-64.
Fisher, R.A. 1922. On the dominance ratio. *Proc. Roy. Soc. Edinb.* 52, 321-341.
Fisher, R.A. 1930. *The Genetical Theory of Natural Selection*. Clarendon Press, Oxford.
Fisher, R.A. 1958. *The Genetical Theory of Natural Selection*. 2nd ed. Dover, New York.
Gantmacher, F.R. 1959. *The Theory of Matrices*. 2 vol. Chelsea, New York.
Geiringer, H. 1944. On the probability theory of linkage in Mendelian heredity. *Ann. Math. Stat.* 15, 25-57.

Geiringer, H. 1948. On the mathematics of random mating in case of different recombination values for males and females. *Genetics* 33, 548-564.

Haldane, J.B.S. 1924. A mathematical theory of natural and artificial selection. Part I. *Trans. Camb. Phil. Soc.* 23, 19-41.

Haldane, J.B.S. 1924a. A mathematical theory of natural and artificial selection. Part II. *Proc. Camb. Phil. Soc., Biol. Sci.* 1, 158-163.

Haldane, J.B.S. 1926. A mathematical theory of natural and artificial selection. Part III. *Proc. Camb. Phil. Soc.* 23, 363-372.

Haldane, J.B.S. 1927. A mathematical theory of natural and artificial selection. Part IV. *Proc. Camb. Phil. Soc.* 23, 607-615.

Haldane, J.B.S. 1927a. A mathematical theory of natural and artificial selection. Part V. Selection and mutation. *Proc. Camb. Phil. Soc.* 23, 838-844.

Haldane, J.B.S. 1930. A mathematical theory of natural and artificial selection. Part VI. Isolation. *Proc. Camb. Phil. Soc.* 26, 220-230.

Haldane, J.B.S. 1931. A mathematical theory of natural and artificial selection. Part VII. Selection intensity as a function of mortality rate. *Proc. Camb. Phil. Soc.* 27, 131-136.

Haldane, J.B.S. 1931a. A mathematical theory of natural and artificial selection. Part VIII. Metastable populations. *Proc. Camb. Phil. Soc.* 27, 137-142.

Haldane, J.B.S. 1932. A mathematical theory of natural and artificial selection. Part IX. Rapid selection. *Proc. Camb. Phil. Soc.* 28, 244-248.

Haldane, J.B.S. 1934. A mathematical theory of natural and artificial selection. Part X. Some theorems on artificial selection. *Genetics* 19, 412-429.

Haldane, J.B.S. 1935. The rate of spontaneous mutation of a human gene. *J. Genet.* 31, 317-326.

Haldane, J.B.S., and Jayakar, S.D. 1963. Polymorphism due to selection of varying direction. *J. Genet.* 58, 237-242.

Hardy, G.H. 1908. Mendelian proportions in a mixed population. *Science* 28, 49-50.

Hartl, D.L. 1970. A mathematical model for recessive lethal segregation distorters with differential viabilities in the sexes. *Genetics* 66, 147-163.

Hartl, D.L. 1970a. Analysis of a general population genetic model of meiotic drive. *Evolution* 24, 538-545.

Hartl, D.L. 1970b. Population consequences of non-Mendelian segregation among multiple alleles. *Evolution* 24, 415-423.

Hoekstra, R.F. 1975. A deterministic model of cyclical selection. *Genet. Res.* 25, 1-15.

Hughes, P.J., and Seneta, E. 1975. Selection equilibria in a multi-allele single-locus setting. *Heredity* 35, 185-194.

Jennings, H.S. 1916. The numerical results of diverse systems of breeding. *Genetics* 1, 53-89.

Jennings, H.S. 1917. The numerical results of diverse systems of breeding, with respect to two pairs of characters, linked or independent, with special relation to the effects of linkage. *Genetics* 2, 97-154.

Karlin, S. 1968. Equilibrium behavior of population genetic models with non-random mating. *J. Appl. Prob.* 5, 231-313, 487-566. Reprinted by Gordon and Breach, New York (1969).

Karlin, S. 1975. General two-locus selection models: some objectives, results, and interpretations. *Theor. Pop. Biol.* 7, 364-398.

Karlin, S. 1976. Population subdivision and selection-migration interaction. In *Proc. Int. Conf. Pop. Genet. Ecol.* (ed. S. Karlin and E. Nevo). Academic Press, New York.

Karlin, S., and Feldman, M.W. 1968. Further analysis of negative assortative mating. *Genetics* 59, 117-136.

Karlin, S., and Levikson, B. 1974. Temporal fluctuations in selection intensities: case of small population size. *Theor. Pop. Biol.* 6, 383-412.

Karlin, S., and Lieberman, U. 1974. Random temporal variation in selection intensities: case of large population size. *Theor. Pop. Biol.* 6, 355-382.

Karlin, S., and McGregor, J. 1972. Application of method of small parameters to multi-niche population genetic models. *Theor. Pop. Biol.* 3, 186-209.

Karlin, S., and McGregor, J. 1972a. Polymorphisms for genetic and ecological systems with weak coupling. *Theor. Pop. Biol.* 3, 210-238.

Karlin, S., and Richter-Dyn, N. 1976. Some theoretical analyses of migration-selection interaction in a cline: a generalized two-range environment. In *Proc. Int. Conf. Pop. Genet. Ecol.* (ed. S. Karlin and E. Nevo). Academic Press, New York.

Kimura, M. 1956. Rules for testing stability of a selective polymorphism. *Proc. Nat. Acad. Sci. USA* 42, 336-340.

Kimura, M. 1956a. A model of a genetic system which leads to closer linkage by natural selection. *Evolution* 10, 278-287.

Kimura, M. 1958. On the change of population fitness by natural selection. *Heredity* **12**, 145-167.

Kimura, M. 1960. *Outline of Population Genetics*. Baifukan, Tokyo (in Japanese).

Kimura, M. 1965. Attainment of quasi-linkage equilibrium when gene frequencies are changing by natural selection. *Genetics* **52**, 875-890.

Kimura, M. 1968. Evolutionary rate at the molecular level. *Nature* **217**, 624-626.

Kimura, M., and Ohta, T. 1971. *Theoretical Aspects of Population Genetics*. Princeton University Press, Princeton.

King, J.L. 1965. The effect of litter culling--of family planning--on the rate of natural selection. *Genetics* **51**, 425-429.

King, J.L., and Jukes, T.H. 1969. Non-Darwinian evolution. *Science* **164**, 788-798.

Kingman, J.F.C. 1961. A matrix inequality. *Quart. J. Math.* **12**, 78-80.

Kingman, J.F.C. 1961a. A mathematical problem in population genetics. *Proc. Camb. Phil. Soc.* **57**, 574-582.

Levene, H. 1953. Genetic equilibrium when more than one ecological niche is available. *Am. Nat.* **87**, 311-313.

Levy, J. 1976. A review of evidence for a genetic component in the determination of handedness. *Behav. Genet.* **6**, 429-453.

Levy, J., and Nagylaki, T. 1972. A model for the genetics of handedness. *Genetics* **72**, 117-128.

Levy, J., and Reid, M. 1976. Variations in writing posture and cerebral organization. *Science* **194**, 337-339.

Levy, J., and Reid, M. 1977. Variations in cerebral organization as a function of handedness, hand posture in writing, and sex. Submitted to *J. Exp. Psychol.: General*.

Lewontin, R.C. 1958. A general method for investigating the equilibrium of gene frequency in a population. *Genetics* **43**, 421-433.

Lewontin, R.C., and Kojima, K. 1960. The evolutionary dynamics of complex polymorphisms. *Evolution* **14**, 458-472.

Li, C.C. 1955. *Population Genetics*. The University of Chicago Press, Chicago. 2nd ed.: Boxwood Press, Pacific Grove, Calif. (1976).

Malécot, G. 1948. *Les mathêmatiques de l'hérédité*. Masson, Paris. Extended translation: *The Mathematics of Heredity*. W.H. Freeman, San Francisco (1969).

Mandel, S.P.H. 1959. The stability of a multiple allelic system. *Heredity* **13**, 289-302.

Mandel, S.P.H. 1970. The equivalence of different sets of stability conditions for multiple allelic systems. *Biometrics* **26**, 840-845.

Maynard Smith, J. 1966. Sympatric speciation. *Am. Nat.* **100**, 637-650.

Maynard Smith, J. 1970. Genetic polymorphism in a varied environment. *Am. Nat.* **104**, 487-490.

Moran, P.A.P. 1962. *The Statistical Processes of Evolutionary Theory.* Clarendon Press, Oxford.

Moran, P.A.P. 1967. Unsolved problems in evolutionary theory. *Proc. Fifth Berkeley Symp. Math. Stat. Prob.* **4**, 457-480.

Mulholland, H.P., and Smith, C.A.B. 1959. An inequality arising in genetical theory. *Am. Math. Mon.* **66**, 673-683.

Nagylaki, T. 1974. Continuous selective models with mutation and migration. *Theor. Pop. Biol.* **5**, 284-295.

Nagylaki, T. 1975. Polymorphisms in cyclically-varying environments. *Heredity* **35**, 67-74.

Nagylaki, T. 1975a. The deterministic behavior of self-incompatibility alleles. *Genetics* **79**, 545-550.

Nagylaki, T. 1975b. Conditions for the existence of clines. *Genetics* **80**, 595-615.

Nagylaki, T. 1975c. A continuous selective model for an X-linked locus. *Heredity* **34**, 273-278.

Nagylaki, T. 1976. The evolution of one- and two-locus systems. *Genetics* **83**, 583-600.

Nagylaki, T. 1976a. A model for the evolution of self-fertilization and vegetative reproduction. *J. Theor. Biol.* **58**, 55-58.

Nagylaki, T. 1976b. Dispersion-selection balance in localised plant populations. *Heredity* **37**, 59-67.

Nagylaki, T. 1977. The geographical structure of populations. In *Studies in Mathematical Biology* (ed. S. Levin). Math. Assoc. Amer., Washington, D.C.

Nagylaki, T. 1977a. The evolution of one- and two-locus systems. II. *Genetics*, in press.

Nagylaki, T., and Crow, J.F. 1974. Continuous selective models. *Theor. Pop. Biol.* **5**, 257-283.

Nagylaki, T., and Levy, J. 1973. "The sound of one paw clapping" isn't sound. *Behav. Genet.* **3**, 279-292.

Nei, M. 1971. Fertility excess necessary for gene substitution in regulated populations. *Genetics* **68**, 169-184.

Nei, M. 1975. *Molecular Population Genetics and Evolution.* North-Holland, Amsterdam.

Norton, H.T.J. 1928. Natural selection and Mendelian variation. *Proc. Lond. Math. Soc.* **28**, 1-45.

O'Donald, P. 1960. Assortative mating in a population in which two alleles are segregating. *Heredity* **15**, 389-396.

Palm, G. 1974. On the selection model for a sex-linked locus. *Math. Biol.* **1**, 47-50.

Penrose, L.S. 1949. The meaning of "fitness" in human populations. *Ann. Eugen.* **14**, 301-304.

Penrose, L.S., Maynard Smith, S., and Sprott, D.A. 1956. On the stability of allelic systems, with special reference to haemoglobins A, S, and C. *Ann. Hum. Genet.* **21**, 90-93.

Prout, T. 1968. Sufficient conditions for multiple niche polymorphism. *Am. Nat.* **102**, 493-496.

Robbins, R.B. 1918. Some applications of mathematics to breeding problems. II. *Genetics* **3**, 73-92.

Robbins, R.B. 1918a. Some applications of mathematics to breeding problems. III. *Genetics* **3**, 375-389.

Scheuer, P.A.G., and Mandel, S.P.H. 1959. An inequality in population genetics. *Heredity* **13**, 519-524.

Scudo, F.M., and Karlin, S. 1969. Assortative mating based on phenotype. I. Two alleles with dominance. *Genetics* **63**, 479-498.

Searle, A.G. 1974. Mutation induction in mice. *Adv. Rad. Biol.* **4**, 131-207.

Snyder, L.H. 1932. Studies in human inheritance. IX. The inheritance of taste deficiency in man. *Ohio J. Sci.* **32**, 436-440.

Sprott, D.A. 1957. The stability of a sex-linked allelic system. *Ann. Hum. Genet.* **22**, 1-6.

Srb, A.M., Owen, R.D., and Edgar, R.S. 1965. *General Genetics*. 2nd ed. W.H. Freeman, San Francisco.

Trankell, A. 1955. Aspects of genetics in psychology. *Am. J. Hum. Genet.* **7**, 264-276.

Vogel, F. 1970. Spontaneous mutation in man. In *Chemical Mutagenesis in Mammals and Man* (ed. F. Vogel and G. Röhrborn). Springer-Verlag, Berlin.

Wahlund, S. 1928. Zusammensetzung von Populationen und Korrelationserscheinungen vom Standpunkt der Vererbungslehre aus betrachtet. *Hereditas* **11**, 65-106.

Wallace, B. 1968. *Topics in Population Genetics*. W.W. Norton, New York.

Weinberg, W. 1908. Über den Nachweis der Vererbung beim Menschen. *Jahresh. Verein. f. vaterl. Naturk. Württem.* **64**, 368-382.

Weinberg, W. 1909. Über Vererbungsgesetze beim Menschen. *Zeit. ind. Abst. Vererb.* **1**, 277-330, 377-392, 440-460; **2**, 276-330.

Wright, S. 1921. Systems of mating. *Genetics* 6, 111-178.

Wright, S. 1931. Evolution in Mendelian populations. *Genetics* 16, 97-159.

Wright, S. 1942. Statistical genetics and evolution. *Bull. Am. Math. Soc.* 48, 223-246.

Wright, S. 1955. Classification of the factors of evolution. *Cold Spring Harbor Symp. Quant. Biol.* 20, 16-24.

Wright, S. 1969. *Evolution and the Genetics of Populations*. Vol. II. The University of Chicago Press, Chicago.

Wright, S. 1970. Random drift and the shifting balance theory of evolution. In *Mathematical Topics in Population Genetics* (ed. K. Kojima). Springer-Verlag, Berlin.

SUBJECT INDEX

ABO blood group, 37-38
Age
    distribution, 14-18, 31-32, 78-80
    independence, 17-18, 80-81
    structure, 14-18, 31-32, 79-82
Additive genetic variance, *see* Variance
Allele, 4
    frequency, *see* Gene frequency
Amino acid, 3
Analysis, 7-8
    complete, *see* global
    global, 7, 57, 92-93, 96-98, 104-106, 119
    local, 7-8, 96-98, 113
Asexual populations, 5-32
Assortative mating, 1, 95-96, 101-107
Autosome, 40
Average
    effect, 89
    excess, 90

Base, 3
Birth rate, *see* Fertility
Blood groups, 37-38
    ABO, 37-38
    MN, 37
Breeding ratios, 38-39, 44-45

Chromosome, 4, 40-41, 43
Continuous time, *see* Selection
Coupling, 43
Convergence, 10, 12-13, 57-60, 92-94, 96-98, 113
    algebraic, 13, 58-60
    geometric, 10, 12-13, 58, 92-94
Covariance, 91
    genic, 91
Crossing over, 43, 50

Death rate, *see* Mortality
de Finetti diagram, 47-48
Deme, 132
Deoxyribonucleic acid, 3
Difference equations
  first order
    linear, homogeneous, 5-7, 96-98
    linear, inhomogeneous, 10, 99
    nonlinear, 11-13, 57, 92-93, 104-106, 113, 119
  higher order, 99
  linear fractional, 11-13
  linearization, 113
  matrix, 96-99
  multidimensional, 96-99
Dioecious population, 33, 36, 50
Diploids, 4
Disassortative mating, 1, 95-96, 112-122
Discrete generations, 5
Dispersion, *see* Migration
DNA, 3
Dominance, 37, 55
  degree of, 55
  deviation, 89
  overdominance, 55-60
  underdominance, 55-60
  variance, *see* Variance
Dominant allele, 37, 55
Drift, genetic, *see* Random genetic drift
Duplicate genes, 43-46

Environment, 47, 55, 70-71
  cyclic, 70-71
  variable, 68-71
Epistatic deviation, 174-190
Epistasis, 174, 190
Equilibrium, 7-8; *see also* Analysis, Difference equations, *and* Polymorphism
  asymptotically stable, 8
  stable, 8
  unstable, 8
Events, 33
  independent, 33

mutually exclusive, 33
Evolution, 2
Fisherian, 2
neutral, 2-3
non-Darwinian, 2-3
Wrightian, 2
Expectation, 46

Fertilization, 4
Fertility, 5, 15, 51-55, 79-83, 95-96, 152-153, 163
multiplicative, 54, 155, 163
nonmultiplicative, 107-112
Fitness, 1, 5, 14-18, 51-55, 79-83, 153, 163, 165-168, 182-184
continuous time, 14-18, 79-83, 182-184
mean, *see* Mean fitness
multiplicative, 54, 92
scaling, 6, 55
surface, 61, 64-66, 189
Fixation of allele, 7
Frequency dependence, *see* Selection
Frobenius' theorem, 134
Fundamental Theorem of Natural Selection, 2, 8-9, 61-64, 84-92, 169-190
continuous time, 19-20, 84-92, 184-188
two loci, 169-190

Gamete, 4, 42-43, 165, 167
Gametic
dispersion, 133
frequency, 42-43, 165, 167
phase equilibrium, 43
selection, 51, 54-55
Gene, 4
duplicate, 44-46
frequency, 6, 51-55
Genetic
code, 3
drift, *see* Random genetic drift
variance, *see* Variance
Genic variance, *see* Variance
Genotypic frequency, 33-34, 51-55, 165, 167
Geographical structure, 3, 124-150

Handedness, 47-50
Haploids, 4-32
Hard selection, 132, 136-138, 140, 144-146
Hardy-Weinberg law, 33-50, 51-55, 84-88, 151-156, 165-167, 184-187
    continuous time, 84-88, 184-187
    dioecious population, 36, 163-164
    monoecious population, 33-36, 47-48, 51-55, 84-88
    multiple loci, 165-167
    two loci, 42-50, 165-167, 184-187
    X-linked locus, 40-42, 151-156
Heterosis, *see* Overdominance
Heterozygosity, 47
Heterozygote, 34
Homozygosity, 46
Homozygote, 34

Inbreeding, 1, 3, 66-68, 94, 96-102, 122-123
    coefficient, 66, 94
    partial selfing, 101-102, 122-123
    selfing with selection, 96-100
    selection with, 66-68, 94, 96-100
Independent loci, 43
Intra-family selection, 71-73
Island model, 14, 124-130

Jensen's inequality, 61-62
Juvenile dispersion, 141-142

Kronecker delta, 11

Lagrange multipliers, 61
Lethal allele, 6, 55, 92, 163
Levene model, 142-146
Life table, 14-18, 79-80
Linear fractional transformation, 11-13
Linkage, 42-46, 165-190
    disequilibrium, 43, 168-190
    equilibrium, 43, 168
Locus, 4
Logistic equation, 22-23
Loss of allele, 7

Malthusian parameter, 18, 79-83, 182-184
Mapping, *see* Difference equations, first order
Maternal inheritance, 73-74
Mating, 1, 33-42, 51-55, 79-84, 95-96, 152-155, 165-167, 182-184
    assortative, *see* Assortative mating
    disassortative, *see* Disassortative mating
    frequency, 35-36, 41, 51-55, 79-84, 95-96, 152, 165-167, 182
    random, *see* Panmixia
Matrix
    irreducible, 134-135
    maximal
        eigenvalue, 134-135
        eigenvector, 134-135
    nonnegative, 134-135
    positive, 138
Mean, 69
    arithmetic, 69
    geometric, 69
    harmonic, 143-144, 150
Mean fitness, 6, 8-9, 53-55, 61-66, 88-92, 142-145, 169-190
    continuous time, 19-20, 88-92, 184-188
    Levene model, 142-145
    stationarity, 61, 189
    two loci, 169-190
Meiosis, 4
Meiotic drive, 74-75
Mendel's Law of Segregation, 34
Migration, 1, 14, 124-150
    gametic, 133
    haploids, 14, 133
    island model, 14, 124-130
    juvenile, 141-142
    Levene model, 142-146
    matrix, 130-131
    multiple niches, 130-142
    random outbreeding and site homing, 134, 140, 147-148
    selection and, *see* Selection
    two niches, 146-150
    weak, 135-137
    zygotic, *see* juvenile
MN blood group, 37
Monoecious population, 33

Mortality, 15, 80
Multiple loci, 165-167
Mutation, 1, 9-14, 29-31, 75-79, 151-157, 160-163
    rate, 9
    selection and, *see* Selection

Nonoverlapping generations, 5
Nucleotide, 3

O, *o*, 63
Ordering of frequencies, 33-42
Overdominance, 55-60
    inbreeding, 68
    intra-family selection, 71-73
    Levene model, 144-145
    maternal inheritance, 73-74
    variable environments, 68-71
    two loci, 177-181
    X-linkage, 158-160
Overlapping generations, *see* Selection *and* Age

Panmixia, 1, 33-50, 54, 84, 155, 165-167, 182-184
    continuous time, 84, 182-184
    dioecious population, 36, 50, 163-164
    monoecious population, 33-36, 47-48, 51-55, 84
    multiple loci, 165-167
    two loci, 42-50, 167, 182-184
    X-linkage, 40-42, 49, 155
Penetrance, 47
Perron's theorem, 138
Phenotype, 37
Phenylketonuria (PKU), 37
Phenylthiocarbamide (PTC) 37
PKU, 37
Pollen elimination, 115-122
Polymorphism, 10
    assortative mating, 102-107
    competitive, 23-29
    cyclic environment, 70-71
    frequency-dependent, 21-22, 31, 93-94
    inbreeding and selection, 66-68
    intra-family selection, 71-73

maternal inheritance, 73-74
migration-selection, 124-150
mutation, 10, 31, 156-157
multiallelic, 60-66
mutation-selection, 10-14, 31, 75-79, 94, 160-163
neutral, 33-50
overdominant, 55-60
partial selfing, 101-102
protected, *see* Protected polymorphism
self-incompatibility, 112-122
selfing with selection, 96-100
two-locus, 180-181, 189-190
underdominant, 55-60
variable environment, 68-71
X-linked, 157-160, 163
Population
regulation, 22-23, 130-134
subdivision, 46-47, 124-150
Position effect, 167
Principal minor, 61
Probability, 33
conditional, 33
joint, 33
Protected polymorphism, 68, 134-135, 138-139
frequency-dependent, 93-94
intra-family selection, 73
Levene model, 142-146
maternal inheritance, 74
migration-selection, 130-150
hard selection, 132, 136-138, 140, 144-146
juvenile dispersion, 141-142
recessive allele, 138-142, 144, 148-150
soft selection, 130-150
multiple niches, 130-142
hard selection, 132, 136-138, 140
juvenile dispersion, 141-142
recessive allele, 138-142
soft selection, 130-142
self-incompatibility alleles, 114-115
two niches, 146-150
variable environments, 68-71
X-linkage, 157-160

Protection of rare alleles, 65-66, 134-136, 138-139, 177-180; *see also* Protected polymorphism
Protein, 3
PTC, 37

Quantitative genetics, 3

Random genetic drift, 1, 3
Random mating, *see* Panmixia
Rate of convergence, *see* Convergence
Recessive allele, 37, 55
Recombination, 43, 50
    frequency, 43, 50
Recursion relations, *see* Difference equations
Regulation of population, 22-23, 130-134
Repulsion, 43

Scaling of fitnesses, 6, 55
Segregation distortion, 74-75
Selection, 1
    age structure with, *see* continuous time
    assortative mating and, 102-104
    autosomal locus, 51-94, 163-164
    change in character under, 88-92
    coefficient, 6, 55, 63-64
    continuous time, 14-32, 79-92, 182-188
    cyclic, 70-71
    fluctuating, 3
    frequency-dependent, 21-22, 31, 93-94
    gametic, 51, 54-55
    haploid, 5-32
    hard, 132, 136-138, 140, 144-146
    inbreeding with, 66-68, 94, 96-100
    intensity, 6, 55, 63-64
    intra-family, 71-73
    island model, 14, 124-130
    Levene model, 142-146
    maternal inheritance, 73-74
    meiotic drive, 74-75
    migration with, 14, 124-150
        island model, 14, 124-130
        Levene model, 142-146

multiple niches, 130-142
random outbreeding and site homing, 147-148
two niches, 146-150
multiple loci, 165-167
multiple niches, 130-142
hard selection, 132, 136-138, 140
juvenile dispersion, 141-142
recessive allele, 138-142
soft selection, 130-142
mutation with
diploids, 75-79, 94
haploids, 9-14, 29-31
X-linkage, 151-157, 160-163
nonmultiplicative fertilities, 107-112
overlapping generations, *see* continuous time
random outbreeding and site homing, 134, 140, 147-148
selfing with, 96-100
sex-linkage, *see X*-linkage *and Y*-linkage
soft, 132-150
time-dependent, 68-71
two loci, 165-190
continuous time, 182-188
multiple alleles, 167-177, 182-190
two alleles, 177-181, 190
two niches, 146-150
weak, 9, 63-64, 84-92, 169-177, 184-188
X-linkage, 151-164
multiple alleles, 151-157
mutation-selection balance, 160-163
two alleles, 157-160, 163
Y-linkage, 5, 40-41
Self-fertilization, 96-102, 122-123
Self-incompatibility, 112-122
alleles, 112-115
pollen elimination, 115-122
zygote elimination, 115-122
Self-sterility, *see* Self-incompatibility
Sex chromosome, 40-41
Sex-linkage, *see X*-linkage *and Y*-linkage
Sibling distributions, 39-40, 45-50
Sickle-cell anemia, 58, 70
Simplex, 61

Snyder's ratios, 38-39
Soft selection, 132-150
Solution, complete, 7
Stability, *see* Equilibrium
Subdivision of population, 46-47, 124-150

Two loci, 42-46, 49-50, 165-190
    continuous time, 182-188
    multiple alleles, 42-46, 49-50, 167-177
    two alleles, 42-46, 49-50, 177-181

Underdominance, 55-60
Unlinked loci, 43

Variance, 8-9, 46
    additive genetic, *see* genic
    additivity of, 89-90, 174-175
    dominance, 89
    epistatic, 174
    gametic, 173
    genic, 8-9, 63, 89-90, 174-175
    genotypic, 63, 89, 174
    total genetic, *see* genotypic
Variation of parameters, 85
Viability, 5, 51-55, 151, 163
    multiplicative, 52, 92

Wahlund's principle, 46-47

X chromosome, 40-41
X-linkage, 40-42, 151-164

Y chromosome, 40-41
Y-linkage, 5, 40-41

Zygote, 4, 51
    elimination, 115-122
    migration, *see* Juvenile dispersion